CHICAGO PUBLIC LIBRARY
NORTH PULASKI BRANCH
4300 W. NORTH AVE.

Delmar Learning's Test Preparation Series

Medium/Heavy Duty Truck Test

Drive Train (Test T3)

3rd Edition

THOMSON

DELMAR LEARNING

Australia Canada Mexico Singapore Spain United Kingdom United States

THOMSON
DELMAR LEARNING

Delmar Learning's ASE Test Preparation Series

ASE Medium/Heavy Duty Truck Test T3 (Drivetrain), 3e

Vice President, Technology and Trades SBU:
Alar Elken

Executive Director, Professional Business Unit:
Greg Clayton

Product Development Manager:
Timothy Waters

Development:
Kristen Shenfield

Channel Manager:
Beth A. Lutz

Marketing Specialist:
Brian McGrath

Production Director:
Mary Ellen Black

Production Manager:
Larry Main

Production Editor:
Elizabeth Hough

Editorial Assistant:
Kristen Shenfield

Cover Designer:
Michael Egan

COPYRIGHT © 2004 by Delmar Learning, a division of Thomson Learning, Inc. Thomson Learning™ is a trademark used herein under license.

Printed in Canada
2 3 4 5 XX 05

For more information contact
Delmar Learning
Executive Woods
5 Maxwell Drive, PO Box 8007,
Clifton Park, NY 12065-8007
Or find us on the World Wide Web at:
www.delmarlearning.com
or www.trainingbay.com

ALL RIGHTS RESERVED. No part of this work covered by the copyright hereon may be reproduced in any form or by any means—graphic, electronic, or mechanical, including photocopying, recording, taping, Web distribution or information storage and retrieval systems—without written permission of the publisher.

For permission to use material from this text or product, contact us by
Tel.   (800) 730-2214
Fax   (800) 730-2215

www.thomsonrights.com

ISBN: 1-4018-2033-6

NOTICE TO THE READER

Publisher does not warrant or guarantee any of the products described herein or perform any independent analysis in connection with any of the product information contained herein. Publisher does not assume, and expressly disclaims, any obligation to obtain and include information other than that provided to it by the manufacturer.

The reader is expressly warned to consider and adopt all safety precautions that might be indicated by the activities herein and to avoid all potential hazards. By following the instructions contained herein, the reader willingly assumes all risks in connection with such instructions.

The Publisher makes no representation or warranties of any kind, including but not limited to, the warranties of fitness for particular purpose or merchantability, nor are any such representations implied with respect to the material set forth herein, and the publisher takes no responsibility with respect to such material. The publisher shall not be liable for any special, consequential, or exemplary damages resulting, in whole or part, from the readers' use of, or reliance upon, this material.

Contents

R0407690533

CHIGAGO PUBLIC LIBRARY
NORTH PULASKI BRANCH
4300 W. NORTH AVE.

Section 4 Overview of the Task List

Section 5 Sample Test for Practice

Section 6 Additional Test Questions for Practice

Section 7 Appendices

Preface

Delmar Learning is very pleased that you have chosen our ASE Test Preparation Series to prepare yourself for the truck ASE Examinations. These guides are available for all of the truck areas that include T1 through T8. These guides are designed to introduce you to the Task List for the test you are preparing to take, give you an understanding of what you are expected to be able to do in each task, and take you through sample test questions formatted in the same way the ASE tests are structured. If you have a working knowledge of the discipline you are testing for, you will find the Delmar ASE Test Preparation Series to be an excellent way to understand the "must know" items to pass the test. These books are not textbooks. Their objective is to prepare the technician who has the requisite experience and schooling to challenge ASE testing. It cannot replace the hands-on experience or the theoretical knowledge required by ASE to master vehicle repair technology. If you are unable to understand more than a few of the questions and their explanations in this book, it could be that you require either more shop-floor experience or further study. Some textbooks that can assist you with further study are listed on the rear cover of this book.

Each book begins with an item-by-item overview of the ASE Task List with explanations of the minimum knowledge you must possess to answer questions related to the task. Following that there are 2 sets of sample questions followed by an answer key to each test and an explanation of the answers to each question. A few of the questions are not strictly ASE format but were included because they help to teach a critical concept that will appear on the test. We suggest that you read the complete Task List Overview before taking the first sample test. After taking the first test, score yourself and read the explanation to any questions that you were not sure about, including the questions you answered correctly. Each test question has a reference back to the related task or tasks that it covers. This will help you to go back and re-read any area of the task list that you are having trouble with. Once you are satisfied that you have all of your questions answered from the first sample test take the second one and check it. If you pass these tests, you will do well on the ASE test.

Our Commitment to Excellence

The 3rd edition of Delmar Learning's ASE Test Preparation Series has been through a major revision with extensive updates to the ASE's Task Lists, test questions, and accuracy. Delmar Learning has sought out the best technicians in the country to help with the updating and revision of each of the books in the series.

About the Revision Author

Martin Restoule is a college professor in Ottawa, Ontario, Canada. Martin is a certified Inter-provincial Automotive Service Technician and Truck and Coach Technician with 25 years experience. Martin has taught these trades for the past 15 years and has taken part in numerous updating courses over this time. He has been very active in a number of trade-related associations and committees and has taken part in many Truck and Coach and Automotive curriculum development committees. Martin also enjoys serving his community as a Senator in a local service club.

About the Series Editor

To promote consistency throughout the series, a series advisor took on the task of reading, editing, and helping our experts give each book the highest level of accuracy possible. Sean Bennett has served in the role of Series Advisor for the 3rd edition of the ASE Test Preparation Series. Sean Bennett has an industry background with a major truck corporation in Allentown, PA, and currently coordinates truck technology training for a truck company in Toronto. He has held ASE Master (Truck) Certification and taught at both college and corporate training levels. Additionally, he is the author of a number of Delmar Learning's publications including *Truck Engine, Fuel and Computerized Management Systems, Heavy Duty Truck Systems, Truck Diesel Engines, and ASE Test Preparation for School Bus books S2, S4, and S5*. Sean is also the revision author for book T6 in this series. He has made a lifelong commitment to adult education and believes that programs emphasizing performance-based learning outcomes are critical for student success in transportation technology education.

Thanks for choosing Delmar Learning's ASE Test Preparation Series. All of the writers, editors, Delmar Learning staff, and myself have worked very hard to make this series second to none. I know you are going to find this book accurate and easy to work with. It is our objective to constantly improve our product at Delmar Learning by responding to feedback. If you have any questions concerning the books in this series you can email me at truckexpert@trainingbay.com.

Sean Bennett
Series Advisor

The History of ASE

History

Originally known as The National Institute for Automotive Service Excellence (NIASE), today's ASE was founded in 1972 as a nonprofit, independent entity dedicated to improving the quality of automotive service and repair through the voluntary testing and certification of automotive technicians. Until that time, consumers had no way of distinguishing between competent and incompetent automotive mechanics. In the mid-1960s and early 1970s, efforts were made by several automotive industry affiliated associations to respond to this need. Though the associations were nonprofit, many regarded certification test fees merely as a means of raising additional operating capital. Also, some associations, having a vested interest, produced test scores heavily weighted in the favor of its members.

From these efforts a new independent, nonprofit association, the National Institute for Automotive Service Excellence (NIASE), was established. In early NIASE tests, Mechanic A, Mechanic B type questions were used. Over the years the trend has not changed, but in mid-1984 the term was changed to Technician A, Technician B to better emphasize sophistication of the skills needed to perform successfully in the modern motor vehicle industry. In certain tests the term used is Estimator A/B, Painter A/B, or Parts Specialist A/B. At about that same time, the logo was changed from "The Gear" to "The Blue Seal," and the organization adopted the acronym ASE for Automotive Service Excellence.

ASE

ASE's mission is to improve the quality of vehicle repair and service in the United States through the testing and certification of automotive repair technicians. Prospective candidates register for and take one or more of ASE's many exams.

Upon passing at least one exam and providing proof of two years of related work experience, the technician becomes ASE certified. A technician who passes a series of exams earns ASE Master Technician status. An automobile technician, for example, must pass eight exams for this recognition.

The exams, conducted twice a year at over seven hundred locations around the country, are administered by American College Testing (ACT). They stress real-world diagnostic and repair problems. Though a good knowledge of theory is helpful to the technician in answering many of the questions, there are no questions specifically on theory. Certification is valid for five years. To retain certification, the technician must be retested to renew his or her certificate.

The automotive consumer benefits because ASE certification is a valuable yardstick by which to measure the knowledge and skills of individual technicians, as well as their commitment to their chosen profession. It is also a tribute to the repair facility employing ASE certified technicians. ASE certified technicians are permitted to wear blue and white ASE shoulder insignia, referred to as the "Blue Seal of Excellence," and carry credentials listing their areas of expertise. Often employers display their technicians' credentials in the customer waiting area. Customers look for facilities that display ASE's Blue Seal of Excellence logo on outdoor signs, in the customer waiting area, in the telephone book (Yellow Pages), and in newspaper advertisements.

To become ASE certified, contact:

National Institute for Automotive Service Excellence
101 Blue Seal Drive S.E.
Suite 101
Leesburg, VA 20175
Telephone 703-669-6600
FAX 703-669-6123
www.ase.com

2 Take and Pass Every ASE Test

ASE Testing

Participating in an Automotive Service Excellence (ASE) voluntary certification program gives you a chance to show your customers that you have the "know-how" needed to work on today's modern vehicles. The ASE certification tests allow you to compare your skills and knowledge to the automotive service industry's standards for each specialty area.

If you are the "average" automotive technician taking this test, you are in your mid-thirties and have not attended school for about fifteen years. That means you probably have not taken a test in many years. Some of you, on the other hand, have attended college or taken postsecondary education courses and may be more familiar with taking tests and with test-taking strategies. There is, however, a difference in the ASE test you are preparing to take and the educational tests you may be accustomed to.

Who Writes the Questions?

The questions, written by service industry experts familiar with all aspects of service consulting, are entirely job related. They are designed to test the skills that you need to know to work as a successful technician; theoretical knowledge is not covered.

Each question has its roots in an ASE "item-writing" workshop where service representatives from automobile manufacturers (domestic and import), aftermarket parts and equipment manufacturers, working technicians, and vocational educators meet in a workshop setting to share ideas and translate them into test questions. Each test question written by these experts must survive review by all members of the group. The questions are written to deal with practical application of soft skills and product knowledge experienced by technicians in their day-to-day work.

All questions are pretested and quality-checked on a national sample of technicians. Those questions that meet ASE standards of quality and accuracy are included in the scored sections of the tests; the "rejects" are sent back to the drawing board or discarded altogether.

Each certification test is made up of between forty and eighty multiple-choice questions. The testing sessions are 4 hours and 15 minutes, allowing plenty of time to complete several tests.

Note: Each test could contain additional questions that are included for statistical research purposes only. Your answers to these questions will not affect your score, but since you do not know which ones they are, you should answer all questions in the test. The five-year Recertification Test will cover the same content areas as those listed above. However, the number of questions in each content area of the Recertification Test will be reduced by about one-half.

Objective Tests

A test is called an objective test if the same standards and conditions apply to everyone taking the test and there is only one correct answer to each question. Objective tests primarily measure your ability to recall information. A well-designed objective test can also test your ability to understand, analyze, interpret, and apply your knowledge. Objective tests include true-false, multiple choice, fill in the blank, and matching questions. ASE's tests consist exclusively of four-part multiple-choice objective questions.

Before beginning to take an objective test, quickly look over the test to determine the number of questions, but do not try to read through all of the questions. In an ASE test, there are usually between forty and eighty questions, depending on the subject. Read through each question before marking your answer. Answer the questions in the order they appear on the test. Leave the questions blank that you are not sure of and move on to the next question. You can return to those unanswered questions after you have finished the others. They may be easier to answer at a later time after your mind has had additional time to consider them on a subconscious level. In addition, you might find information in other questions that will help you to answer some of them.

Do not be obsessed by the apparent pattern of responses. For example, do not be influenced by a pattern like **D, C, B, A, D, C, B, A** on an ASE test.

There is also a lot of folk wisdom about taking objective tests. For example, there are those who would advise you to avoid response options that use certain words such as *all, none, always, never, must,* and *only,* to name a few. This, they claim, is because nothing in life is exclusive. They would advise you to choose response options that use words that allow for some exception, such as *sometimes, frequently, rarely, often, usually, seldom,* and *normally.* They would also advise you to avoid the first and last option (A and D) because test writers, they feel, are more comfortable if they put the correct answer in the middle (B and C) of the choices. Another recommendation often offered is to select the option that is either shorter or longer than the other three choices because it is more likely to be correct. Some would advise you to never change an answer since your first intuition is usually correct.

Although there may be a grain of truth in this folk wisdom, ASE test writers try to avoid them and so should you. There are just as many **A** answers as there are **B** answers, just as many **D** answers as **C** answers. As a matter of fact, ASE tries to balance the answers at about 25 percent per choice **A, B, C,** and **D.** There is no intention to use "tricky" words, such as outlined above. Put no credence in the opposing words "sometimes" and "never," for example.

Multiple-choice tests are sometimes challenging because there are often several choices that may seem possible, and it may be difficult to decide on the correct choice. The best strategy, in this case, is to first determine the correct answer before looking at the options. If you see the answer you decided on, you should still examine the options to make sure that none seem more correct than yours. If you do not know or are not sure of the answer, read each option very carefully and try to eliminate those options that you know to be wrong. That way, you can often arrive at the correct choice through a process of elimination.

If you have gone through all of the test and you still do not know the answer to some of the questions, then guess. Yes, guess. You then have at least a 25 percent chance of being correct. If you leave the question blank, you have no chance. In ASE tests, there is no penalty for being wrong.

Preparing for the Exam

The main reason we have included so many sample and practice questions in this guide is, simply, to help you learn what you know and what you don't know. We recommend that you work your way through each question in this book. Before doing this, carefully look through Section 3; it contains a description and explanation of the questions you'll find in an ASE exam.

Once you know what the questions will look like, move to the sample test. After you have answered one of the sample questions (Section 5), read the explanation (Section 7) to the answer for that question. If you don't feel you understand the reasoning for the correct answer, go back and read the overview (Section 4) for the task that is related to that question. If you still don't feel you have a solid understanding of the material, identify a good source of information on the topic, such as a textbook, and do some more studying.

After you have completed the sample test, move to the additional questions (Section 6). This time answer the questions as if you were taking an actual test. Once you have answered all of the questions, grade your results using the answer key in Section 7. For every question that you gave a wrong answer to, study the explanations to the answers and/or the overview of the related task areas.

Here are some basic guidelines to follow while preparing for the exam:

- Focus your studies on those areas you are weak in.
- Be honest with yourself while determining if you understand something.
- Study often but in short periods of time.
- Remove yourself from all distractions while studying.
- Keep in mind the goal of studying is not just to pass the exam, the real goal is to learn!

During the Test

Mark your bubble sheet clearly and accurately. One of the biggest problems an adult faces in test taking, it seems, is in placing an answer in the correct spot on a bubble sheet. Make certain that you mark your answer for, say, question 21, in the space on the bubble sheet designated for the answer for question 21. A correct response in the wrong bubble will probably be wrong. Remember, the answer sheet is machine scored and can only "read" what you have bubbled in. Also, do not bubble in two answers for the same question.

If you finish answering all of the questions on a test ahead of time, go back and review the answers of those questions that you were not sure of. You can often catch careless errors by using the remaining time to review your answers.

At practically every test, some technicians will invariably finish ahead of time and turn their papers in long before the final call. Do not let them distract or intimidate you. Either they knew too little and could not finish the test, or they were very self-confident and thought they knew it all. Perhaps they were trying to impress the proctor or other technicians about how much they know. Often you may hear them later talking about the information they knew all the while but forgot to respond on their answer sheet.

It is not wise to use less than the total amount of time that you are allotted for a test. If there are any doubts, take the time for review. Any product can usually be made better with some additional effort. A test is no exception. It is not necessary to turn in your test paper until you are told to do so.

Your Test Results!

You can gain a better perspective about tests if you know and understand how they are scored. ASE's tests are scored by American College Testing (ACT), a nonpartial, unbiased organization having no vested interest in ASE or in the automotive industry. Each question carries the same weight as any other question. For example, if there are fifty questions, each is worth 2 percent of the total score. The passing grade is 70 percent. That means you must correctly answer thirty-five of the fifty questions to pass the test.

The test results can tell you:

- where your knowledge equals or exceeds that needed for competent performance, or
- where you might need more preparation.

The test results *cannot* tell you:

- how you compare with other technicians, or
- how many questions you answered correctly.

Your ASE test score report will show the number of correct answers you got in each of the content areas. These numbers provide information about your performance in each area of the test. However, because there may be a different number of questions in each area of the test, a high percentage of correct answers in an area with few questions may not offset a low percentage in an area with many questions.

It may be noted that one does not "fail" an ASE test. The technician who does not pass is simply told "More Preparation Needed." Though large differences in percentages may indicate problem areas, it is important to consider how many questions were asked in each area. Since each test evaluates all phases of the work involved in a service specialty, you should be prepared in each area. A low score in one area could keep you from passing an entire test.

There is no such thing as average. You cannot determine your overall test score by adding the percentages given for each task area and dividing by the number of areas. It doesn't work that way because there generally are not the same number of questions in each task area. A task area with twenty questions, for example, counts more toward your total score than a task area with ten questions.

Your test report should give you a good picture of your results and a better understanding of your task areas of strength and weakness.

If you fail to pass the test, you may take it again at any time it is scheduled to be administered. You are the only one who will receive your test score. Test scores will not be given over the telephone by ASE nor will they be released to anyone without your written permission.

3 Are You Sure You're Ready for Test T3?

Pretest

The purpose of this pretest is to determine the amount of review that you may require prior to taking the ASE medium/heavy truck: Drive Train (Test T3). If you answer all of the pretest questions correctly, complete the sample test in Section 5 along with the additional test questions in Section 6.

If two or more of your answers to the pretest questions are wrong, study Section 4: An Overview of the Task List before continuing with the sample test and additional test questions.

The pretest answers and explanations are located at the end of the pretest.

1. Which of the following is the cause of a "grabbing" condition during clutch release?
   A. Worn friction facings on the clutch discs
   B. Release bearing fork wear
   C. Worn release bearing
   D. Warped intermediate plate or pressure plate

2. Under what operating conditions would a pilot bearing make a rattling or growling noise?
   A. Engine idling with the clutch pedal fully depressed and the clutch released
   B. The vehicle decelerating in high gear with the clutch engaged
   C. The vehicle accelerating in low gear with the clutch engaged
   D. Engine idling with the clutch engaged

3. A burnt pressure plate may be caused by all of the following **EXCEPT:**
   A. oil on the clutch friction disc.
   B. not enough clutch pedal free play.
   C. binding linkage.
   D. a damaged pilot bearing.

4. To install a new pull-type clutch a technician will need to do all the following, **EXCEPT:**
   A. align the clutch disc.
   B. adjust the release bearing.
   C. resurface the limited torque clutch brake.
   D. lubricate the pilot bearing.

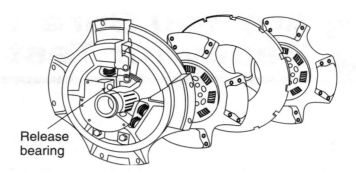

Release
bearing

5. A vehicle with the type of clutch as shown in the figure is brought into the garage because it goes through ceramic friction discs too frequently. What should the technician do?
   A. Instruct the driver on proper operation of the clutch.
   B. Switch to a fiber-type disc.
   C. Replace the pressure plate.
   D. Replace the release bearing.

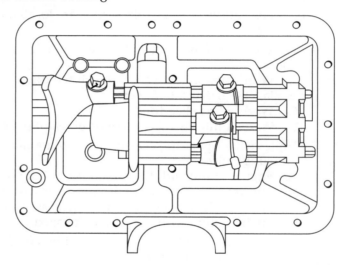

6. When checking the transmission shift cover detents on the shift bar housing as shown in the figure, check for all of the following **EXCEPT:**
   A. worn or oblonged detent recesses.
   B. broken detent springs.
   C. properly lubricated detent spring channels.
   D. a rough or worn detent ball.

7. The LEAST-Likely damage to check for on an input shaft is:
   A. pilot bearing shaft cracking.
   B. gear teeth damage.
   C. input spline damage.
   D. cracking or other fatigue wear to the input shaft spline.

8. Which of the following damage is LEAST-Likely to happen in a two-gear drive combination?
   A. Tooth chipping
   B. Bottoming
   C. Climbing
   D. Spalling

9. Technician A says that pilot bore run out should be measured when a vehicle makes a grinding noise that cannot be identified in any other drive train member. Technician B says that pilot bore run out should be measured only when the pilot bearing is found to be in poor shape. Who is right?
   A. A only
   B. B only
   C. Both A and B
   D. Neither A nor B

10. A technician notices overheated oil coating the seals of the transmission. Technician A says that you must replace all the seals in the transmission. Technician B says that changing to a higher grade of transmission oil may be all that will be necessary. Who is right?
   A. A only
   B. B only
   C. Both A and B
   D. Neither A nor B

11. A tandem axle truck with the power divider lockout engaged has power applied to the forward rear drive axle while no power is applied to the rearward rear drive axle. The Most-Likely cause of the malfunction is:
   A. broken teeth of the forward drive axle ring gear.
   B. broken teeth of the rear drive axle ring gear.
   C. stripped output shaft splines.
   D. damaged interaxle differential.

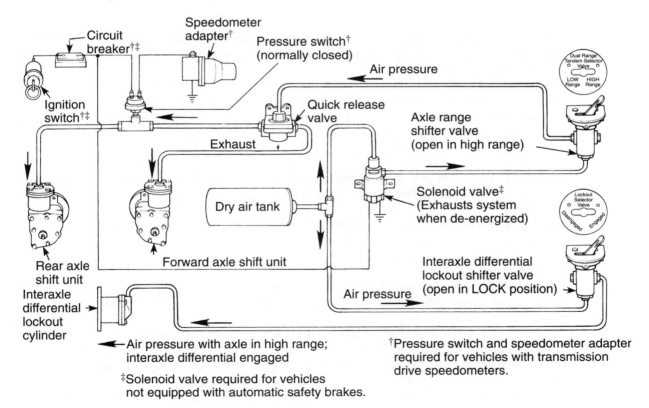

12. This is the axle range and interaxle differential lockout schematic of a vehicle that will not shift from high range to low range. The Most-Likely cause is:
   A. a faulty air compressor.
   B. an air leak at the axle shift unit.
   C. a quick release valve.
   D. a plugged air filter.

Answers to the Test Questions for the Pretest

1. D, 2. A, 3. D, 4. C, 5. A, 6. C, 7. A, 8. B, 9. B, 10. B, 11. C, 12. C

Explanations to the Answers for the Pretest

Question #1
Answer A is wrong; worn clutch friction discs would cause slipping not grabbing.
Both B and C are also wrong; release bearing forks and worn release bearings may cause incomplete clutch disengagement and burning.
Answer D is correct because the warpage of either plate will create inconsistent clutch apply pressures or "grabbing" during release.

Question #2
Answer A is correct. This condition with the engine running and the clutch disengaged allows for a speed difference between the flywheel and the transmission input shaft. This speed difference allows the pilot bearing to operate and generate noise.
Answers B, C, and D all have the clutch engaged, which does not allow for any speed difference.

Question #3
Answer A is wrong because oil on the friction disc provides a lubricant between the friction disc and the pressure plate, which will allow slippage that generates heat and can burn the pressure plate.
Answers B and C prevent complete clutch engagement, which will decrease the pressure plate apply force and allow disc slippage and pressure plate burning.
Answer D is correct because the pilot bearing will have no effect on the friction material or the clamping force of the pressure plate so no burning will take place.

Question #4
Answers A, B, and D are wrong because aligning the clutch disc, adjusting the release bearing, and lubricating the pilot bearing must be performed when installing a new pull-type clutch.
Answer C is correct because you would only resurface the limited torque clutch brake if it had surface damage or unevenness.

Question #5
Answer A is correct. You should instruct the driver to not rest his or her foot on the clutch pedal as this practice reduces the pressure plate apply force, which can cause slippage and disc wear.
Answer B is wrong because fiber discs have a lower coefficient of friction and a shorter life span than ceramic discs.
Answer C is wrong because a new pressure plate will still not apply full pressure against the disc if the driver's foot is resting on the clutch pedal.
Answer D is also wrong because the release bearing is only a part of the release system, which has no effect on the clutch application pressure when adjusted correctly.

Question #6
Answers A, B, and D are wrong because these are all valid checks. Worn or oblong detent recesses may allow the detent balls to wedge in the rail or housing due to the excess wear. Broken detent springs could prevent the detent balls from seating firmly in the recesses. A rough or worn detent ball could wedge in a recess and lock the shift rail in the selected position.
Answer C is correct because detent springs do not require lubrication.

Question #7
Answer A is correct because pilot bearing shaft cracking is the LEAST-Likely damage to occur on the input shaft. Very little movement takes place between the shaft and the pilot bearing.
Answer B is wrong because gear teeth can be damaged by a number of conditions including low lube levels, dirty or contaminated lube, and mating gear tooth damage.
Answer C is wrong because input shaft spline damage can occur during normal clutch disc movement when engaging and disengaging.
Answer D is wrong because driveline and engine rpm irregularities tend to crack and chip input shaft splines.

Question #8
Answer A is wrong because tooth chipping can occur in a two-gear drive combination due to tooth backlash and driveline shocks to the gear set.
Answer B is correct because bottoming is the LEAST-Likely damage to occur. Two gears under load tend to separate from each other.
Answer C is wrong because climbing damage can occur. Helical gears under load tend to climb.
Answer D is wrong because spalling can occur due to fatigue from constant tooth to tooth contact and their meshing action.

Question #9
Answer B is correct. Only Technician B is right. Pilot bearing bore run out should only be checked when the bearing shows signs of damage. Technician A is wrong because this is a rare condition and it is a time-consuming procedure. Other checks should be performed first.

Question #10
Answer B is correct. Technician B is right because changing to a higher grade of transmission oil will provide better lubrication therefore reducing the heat generated. A higher quality, cooler temperature oil will not thin out and creep past the seal lips. Technician A is wrong because changing all of the seals would not correct the cause of the leakage.

Question #11
Answer A is wrong because broken teeth on the forward drive axle ring gear can cause the front axle to be non-powered.
Answer B is wrong because broken teeth on the rear drive axle ring gear would not affect the power flow to the rear axle.
Answer C is correct because stripped output shaft splines would still allow power to reach the front drive axle but the interaxle differential side gear would slip on the output shaft, producing no drive to the rear drive axle.
Answer D is wrong because interaxle differential damage could render both axles powerless.

Question #12

Answers A and B are wrong because a high to low range shift requires air to shift the axle into high range first. A faulty air compressor or an air leak would prevent this.

Answer C is correct because the shift to low range could not take place if the air could not exhaust through the quick release valve.

Answer D is also wrong because the plugged air filter could cause a slow air buildup and high temperature air, which will usually not affect the shift from high to low range.

Types of Questions

ASE certification tests are often thought of as being tricky. They may seem to be tricky if you do not completely understand what is being asked. The following examples will help you recognize certain types of ASE questions and avoid common errors.

Each test is made up of forty to eighty multiple-choice questions. Multiple-choice questions are an efficient way to test knowledge. To answer them correctly, you must think about each choice as a possibility, and then choose the one that best answers the question. To do this, read each word of the question carefully. Do not assume you know what the question is about until you have finished reading it.

Multiple-Choice Questions

One type of multiple-choice question has three wrong answers and one correct answer. The wrong answers, however, may be almost correct, so be careful not to jump at the first answer that seems to be correct. If all the answers seem to be correct, choose the answer that is the most correct. If you readily know the answer, this kind of question does not present a problem. If you are unsure of the answer, analyze the question and the answers. For example:

Question 1:

To remove an automatic transmission oil pump a technician must:

    A. remove the transmission, then the torque converter, then the oil pump.
    B. remove the transmission pan and filter, then remove the oil pump.
    C. remove the transmission pan and filter, then remove the main control valve body, then the oil pump.
    D. remove the transmission, then remove the torque converter, then remove the bell housing, then remove the oil pump.

Analysis:

Answer A is correct. The correct procedure for oil pump removal is to first remove the transmission, then the torque converter, and then remove the oil pump.
Both B and C are wrong because both selections do not require transmission removal and the pump is not accessible through the oil pan.
Answer D is wrong because the bell housing does not have to be removed.

EXCEPT Questions

Another type of question used on ASE tests has answers that are all correct except one. The correct answer for this type of question is the answer that is wrong. The word **EXCEPT** will always be in capital letters. You must identify which of the choices is the wrong answer. If you read quickly through the question, you may overlook what the question is asking and answer the question with the first correct statement. This will make your answer wrong. An example of this type of question and the analysis is as follows:

Question 2:

The following are all reasons for replacement of the pilot bearing **EXCEPT:**

    A. rough action.
    B. binding action.
    C. spalled outer race.
    D. excessive end play.

Analysis:

Answers A, B, and C are wrong because rough, binding actions and spalling are all valid reasons for pilot bearing replacement. All of these conditions suggest deterioration of the bearings or races.
Answer D is correct because excessive end play is the exception. Pilot bearings are not designed to control end play.

Technician A, Technician B Questions

The type of question that is most popularly associated with an ASE test is the "Technician A says . . . Technician B . . . Who is right?" type. In this type of question, you must identify the correct statement or statements. To answer this type of question correctly, you must carefully read each technician's statement and judge it on its own merit to determine if the statement is true.

Typically, this type of question begins with a statement about some analysis or repair procedure. This is followed by two statements about the cause of the problem, proper inspection, identification, or repair choices. You are asked whether the first statement, the second statement, both statements, or neither statement is correct. Analyzing this type of question is a little easier than the other types because there are only two ideas to consider although there are still four choices for an answer.

Technician A . . . Technician B questions are really double-true-false questions. The best way to analyze this kind of question is to consider each technician's statement separately. Ask yourself, is A true or false? Is B true or false? Then select your answer from the four choices. An important point to remember is that an ASE Technician A . . . Technician B question will never have Technician A and B directly disagreeing with each other. That is why you must evaluate each statement independently. An example of this type of question and the analysis of it follows.

Question 3:
A vehicle with wheel end locking hubs will not switch into high. Technician A says to replace the shift collar return spring. Technician B says to clean out and repack the hub. Who is right?
    A. A only
    B. B only
    C. Both A and B
    D. Neither A nor B

Analysis:

Answer A is wrong. It is a good choice because you do replace the shift collar return spring when wheel end locking hubs will not switch into high. Yet, it is wrong because both technicians are right.
Answer B is wrong. It is a good choice because you can clean out and repack the hub to correct this condition. Yet, it is wrong because both technicians are right.
Answer C is correct. Both technicians are right.
Answer D is wrong. Neither technician is wrong.

Questions with a Figure

About 10 percent of ASE questions will have a figure, as shown in the example:

Question 4:

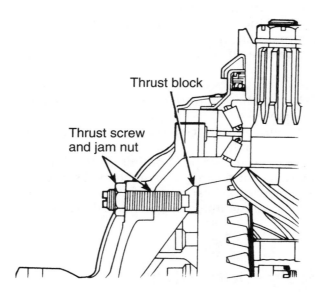

During assembly, a technician notices light score marks on the face of the thrust block as shown in the figure. What does this indicate?

    A. The thrust block must be replaced.
    B. The thrust block is doing its job.
    C. The thrust block is out of adjustment.
    D. The thrust block should be resurfaced.

Analysis:

Answer A is wrong because the marks indicate that the thrust block has done its job and worked correctly.
Answer B is correct because the marks indicate that the thrust block limited ring gear deflection and prevented distortion from excessive thrust.
Answer C is wrong because light score marks indicate that the adjustment was probably correct or there would be either no marks, or excessive burning and scoring on the thrust block.
Answer D is wrong because it is not necessary to resurface the thrust block.

Most-Likely Questions

Most-Likely questions are somewhat difficult because only one choice is correct while the other three choices are nearly correct. An example of a Most-Likely question is as follows:

Question 5:

A tandem axle truck with the power divider lockout engaged has power applied to the forward rear drive axle while no power is applied to the rearward rear drive axle. The Most-Likely cause of the malfunction is:

    A. broken teeth of the forward drive axle ring gear.
    B. broken teeth of the rear drive axle ring gear.
    C. stripped output shaft splines.
    D. damaged interaxle differential.

Analysis:

Answer A is wrong because broken teeth on the forward drive axle ring gear can cause the front axle to be non-powered.

Answer B is wrong because broken teeth on the rear drive axle ring gear would not affect the power flow to the rear axle.

Answer C is correct because stripped output shaft splines would still allow power to reach the front drive axle but the interaxle differential side gear would slip on the output shaft, producing no drive to the rear drive axle.

Answer D is wrong because interaxle differential damage could render both axles powerless.

LEAST-Likely Questions

Notice that in Most-Likely questions there is no capitalization. This is not so with LEAST-Likely type questions. For this type of question, look for the choice that would be the least likely cause of the described situation. Read the entire question carefully before choosing your answer. An example is as follows:

Question 6:

A truck has an extremely hot wheel hub after a test drive. The LEAST-Likely cause would be?

    A. Axle shaft damage
    B. Wheel bearing damage
    C. Air line damage to the brakes
    D. Poor quality lubricant

Analysis:

Answer A is correct. Axle shaft damage is the LEAST-Likely cause of a hot hub because it is most probably the result of the heat.

Answer B is wrong because wheel bearing damage will cause increased friction which will heat the hub and axle housing.

Answer C is wrong because air line damage may allow the brakes to drag or release slowly causing an overheated wheel hub and brake drum assembly.

Answer D is wrong because poor lubricant quality will not provide an adequate film on the axle bearings, increasing friction and heat.

Summary

There are no four-part multiple-choice ASE questions having "none of the above" or "all of the above" choices. ASE does not use other types of questions, such as fill-in-the-blank, completion, true-false, word-matching, or essay. ASE does not require you to draw diagrams or sketches. If a formula or chart is required to answer a question, it is provided for you. There are no ASE questions that require you to use a pocket calculator.

Testing Time Length

An ASE test session is four hours and fifteen minutes. You may attempt from one to a maximum of four tests in one session. It is recommended, however, that no more than a total of 225 questions be attempted at any test session. This will allow for just over one minute for each question.

Visitors are not permitted at any time. If you wish to leave the test room, for any reason, you must first ask permission. If you finish your test early and wish to leave, you are permitted to do so only during specified dismissal periods.

Monitor Your Progress

You should monitor your progress and set an arbitrary limit to how much time you will need for each question. This should be based on the number of questions you are attempting. It is suggested that you wear a watch because some facilities may not have a clock visible to all areas of the room.

Registration

Test centers are assigned on a first-come, first-served basis. To register for an ASE certification test, you should enroll at least six weeks before the scheduled test date. This should provide sufficient time to assure you a spot in the test center. It should also give you enough time for study in preparation for the test. Test sessions are offered by ASE twice each year, in May and November, at over six hundred sites across the United States. Some tests that relate to emission testing also are given in August in several states.

To register, contact Automotive Service Excellence/American College Testing at:

ASE/ACT
P.O. Box 4007
Iowa City, IA 52243
Toll Free: 866-427-3273
www.ase.com

4 Overview of the Task List

Drivetrain (Test T3)

The following section includes the task areas and task lists for this test and a written overview of the topics covered in the test.

The task list describes the actual work you should be able to do as a technician that you will be tested on by the ASE. This is your key to the test and you should review this section carefully. We have based our sample test and additional questions upon these tasks and the overview section will also support your understanding of the task list. ASE advises that the questions on the test may not equal the number of tasks listed; the task lists tell you what ASE expects you to know how to do and be ready to be tested on.

At the end of each question in the Sample Test and Additional Test Questions sections, a letter and number will be used as a reference back to this section for additional study. Note the following example: **A1.**

Task List

A. Clutch Diagnosis and Repair (14 Questions)

Task A1   Diagnose clutch noise, binding, slippage, pulsation, vibration, grabbing, and chatter problems; determine cause of failure and needed repairs.

Example:
1. Which of the following could cause a "grabbing"condition during clutch release?
    A. Worn friction facings on the clutch discs
    B. Release bearing fork wear
    C. Worn release bearing
    D. Warped intermediate plate or pressure plate   (A1)

Question #1
Answer A is wrong; worn clutch friction discs would cause slipping not grabbing.
Both B and C are also wrong; release bearing forks and worn release bearings may cause incomplete clutch disengagement and burning.
Answer D is correct because warpage of either plate will create inconsistent clutch apply pressures or "grabbing" during release.

Task List and Overview

A. Clutch Diagnosis and Repair (14 Questions)

Task A1   **Diagnose clutch noise, binding, slippage, pulsation, vibration, grabbing, and chatter problems; determine cause of failure and needed repairs.**

The most frequent cause of clutch failure is excess heat. The heat generated between the flywheel, driven discs, intermediate plate, and pressure plate may be intense enough to cause the metal to crack or melt and the friction material to be destroyed. Heat or wear is practically nonexistent when the clutch is fully engaged. However, considerable heat can be generated at clutch engagement when the clutch is picking up the load. An improperly adjusted or slipping clutch rapidly generates sufficient heat to self-destruct. Causes of clutch slippage include improper adjustment of an external linkage, a worn or damaged pressure plate, worn clutch disc, grease or oil contamination of the clutch disc, extreme loads, and improper driving techniques.

Task A2   **Inspect, adjust, repair, or replace clutch linkage, cables, levers, brackets, bushings, pivots, springs, and clutch safety switch (includes push and pull-type assemblies); check pedal height and travel.**

"Riding" the clutch pedal is another name for operating the vehicle with the clutch partially engaged. This is very destructive to the clutch, as it permits slippage and generates excessive heat. Riding the clutch also puts constant thrust load on the release bearing, which can thin out the lubricant and cause excessive wear on the pads. Release bearing failures are often the result of this type of driving practice. The best way to determine if clutch disc failure is due to driver error or mechanical failure is to speak with the driver of the vehicle.

The clutch linkage should be adjusted to prevent constant release bearing contact with the clutch release fingers which can produce release bearing wear and may also cause clutch slippage. Too much clearance between the release fingers and the release bearing may not allow full clutch disengagement and clutch brake operation. This adjustment is typically ⅛ inch, which will produce approximately 1½ inch to 2 inches of free travel of the clutch pedal.

Task A3   **Inspect, adjust, repair, or replace hydraulic clutch slave and master cylinders, lines, and hoses; bleed system.**

A typical clutch is controlled and operated by hydraulic fluid pressure and assisted by an air servo cylinder. More specifically, it consists of a master cylinder, hydraulic fluid reservoir, and an air-assisted servo cylinder. These components are all connected using metal and flexible tubes. When the clutch pedal is depressed, the plunger forces the piston in the master cylinder to move forward, causing the hydraulic fluid to act upon the air servo cylinder, which in turn activates the release fork.

Task A4   **Inspect, or remove and install release (throw-out) bearing, sleeve, bushings, springs, housing, levers, release fork, fork pads, fork rollers, shafts, and seals; measure and adjust release (throw-out) bearing position.**

Both push-type and pull-type clutches are disengaged through the movement of a release bearing. The release bearing is a unit within the clutch consisting of bearings that mount on the transmission input shaft sleeve but do not rotate with it. A fork attached to the clutch pedal linkage controls the movement of the release bearing. As the release bearing moves, it forces the pressure plate away from the clutch disc.

Manually adjusted clutches have an adjusting ring that permits the clutch to be manually adjusted to compensate for wearing of the friction linings. The ring is positioned behind the pressure plate and is threaded into the clutch cover. A lock strap or lock plate secures the ring so that it cannot move. The levers are seated in the ring. When the lock strap is removed, the adjusting ring is rotated in the cover so that it moves toward the engine.

Release bearing position is a critical adjustment that ensures proper clutch release and clutch brake operation. The measurement between the release bearing and clutch brake should be ½ inch for most clutch applications. This measurement provides enough release travel to allow for complete clutch disengagement as well as the correct amount of clamping force on the clutch brake when the clutch pedal is fully depressed. When a clutch brake is not installed, the measurement should be ¾ inch between the release bearing and the transmission input shaft bearing retainer.

Task A5   **Inspect, or remove and install single-disc clutch pressure plate and clutch disc; adjust freeplay and release bearing position.**

Single-disc clutch assemblies rely on pressure from the pressure plate on a single clutch disc to transfer full engine torque to the transmission. The friction surfaces on the flywheel and pressure plate must be smooth, free from cracks, hot spots, and contamination. The two types of clutch discs, organic and ceramic, must be inspected for thickness, hot spots, cracking, uneven wear, and oil or grease contamination.

Single disc clutches are available in push or pull types. A push type requires adjustment of only the linkage to provide the adjustment. The linkage should be adjusted to produce approximately ⅛ inch clearance between the release fork and the release bearing to provide the correct pedal free travel. The pull type requires adjustment of the clutch pressure plate for clamp load first, then adjustment of the linkage to produce the correct pedal free travel.

Task A6   **Inspect, or remove and install two-plate clutch pressure plate, clutch disc, intermediate plate; measure and adjust drive pin/lug and/or separator pin clearance; adjust freeplay and release bearing position.**

If the clutch has two driven discs, an intermediate plate or center plate separates the two clutch friction discs. The plate is machined smooth on both sides since it is pressed between two friction surfaces. An intermediate plate increases the torque capacity of the clutch by increasing the friction area, allowing more area for the transfer of torque.

Task A7   **Inspect and replace clutch brake assembly; inspect and replace input shaft and bearing retainer.**

The clutch brake is a circular disc with a friction surface that is mounted on the transmission input spline shaft between the release bearing and the transmission. Its purpose is to slow or stop the transmission-input shaft from rotating in order to allow initial forward or reverse gear engagement without clashing and to keep transmission gear damage to a minimum. Clutch brakes are used only on vehicles with nonsynchronized transmissions.

Task A8   **Inspect, or remove and install self-adjusting/continuous-adjusting clutch assembly; perform initial and/or reset adjustment procedure.**

The wear compensator is a replaceable component that automatically adjusts for facing wear each time the clutch is actuated. Once facing wear exceeds a predetermined amount, the wear compensator allows the adjusting ring to be advanced toward the engine, keeping the pressure plate to clutch disc clearance within proper operating specification. This also keeps free pedal adjustment within specification.

To perform the initial or any manual adjustments necessary during the lifespan of a self-adjusting clutch, the clutch inspection cover must be removed for clutch access and the adjuster mechanism must be rotated to the opening. Remove the adjuster's right mount bolt and loosen the left bolt enough to rotate the assembly upward to disengage it from the adjuster ring. The adjuster ring can now be manually rotated to obtain the proper release bearing position. Once this is achieved, the adjuster can be rotated downward, back into mesh with the adjuster ring and the bolts installed and tightened. Ensure that the adjuster's actuator arm is inserted into the release sleeve retainer because no adjustment will occur if the actuator arm is dislodged.

Task A9   Inspect and replace pilot bearing/bushing.

The pilot bearing or bushing is usually a sealed bearing or a brass bushing. These are responsible for the transmission input shaft alignment with the engine as well as allowing for input shaft rotation in the crankshaft. The pilot bearing or bushing is prone to seizure from its limited movement and from contamination. Misalignment between the engine and transmission can cause premature pilot bearing failure due to the shafts operating at an angle and binding in the bearing. For these reasons and the minor cost of replacement, it is a good practice to replace the pilot bearing or bushing during clutch replacement. An internal puller or a slide hammer can be used to remove either the bearing or bushing types.

Task A10   Inspect flywheel mounting area on crankshaft, rear main oil seal, and measure crankshaft end play; determine needed repairs.

Inspect the crankshaft flywheel mounting surface for any burrs and irregularities, and check all bolt holes for pulled or damaged threads. Examine the crankshaft rear main seal and sealing surface for any signs of damage or oil seepage. Any sign of damage or leaks will require replacement of the seal and sealing surface repair or the installation of a wear sleeve. Crankshaft end play can be checked with a dial indicator. If this check indicates excessive end play, correcting this condition will ensure that no clutch operation problems like excessive flywheel axial movement and oil contamination from the rear main seal will occur. To correct excessive crankshaft end play, the crankshaft thrust bearings must be replaced.

Task A11   Inspect flywheel, starter ring gear, and measure flywheel face and pilot bore run out; determine needed repairs.

The flywheel must be inspected for a number of irregularities. The starter ring gear must be inspected for tooth damage and wear. If the ring gear does show signs of wear it can only be serviced by replacement. The starter drive gear should also be replaced at this time to prevent damage to the new ring gear. The flywheel face and pilot bearing bore should also be checked for run out with a dial indicator. Any run out that exceeds manufacturer's specifications will require machining or replacement of the flywheel. The flywheel face should also be visually inspected for wear conditions such as scoring, bluing, and hot spots.

Task A12   Inspect flywheel housing(s) to transmission housing/engine mating surface(s) and measure flywheel housing face and bore run out; determine needed repairs.

The mating surfaces of the transmission clutch housing and the engine flywheel housing should be inspected for signs of wear or damage. Any appreciable wear on either housing will cause misalignment. To measure either flywheel housing face or bore run out, first attach a dial indicator to the center of the flywheel. Zero the needle on whichever surface is being measured. Turn the flywheel and take special note of the readings, using soapstone or another similar marker to indicate high and low points. As with other runout measurements, subtract the low measurement from the high measurement to get the runout dimension. If runout value is more than specified by the manufacturer, service as necessary.

Task A13   Inspect or replace components of an air-actuated clutch system.

An air actuated clutch control system requires a dedicated and isolated air supply, clutch control valve, clutch brake valve, clutch release air chamber, air applied clutch brake mechanism and regulating valves. The purpose of the isolated supply tank is to provide clutch operation for engine starting in the event of air loss from small leaks in the main air supply circuit or after primary and secondary tank moisture drainage. When the clutch pedal/clutch control valve is depressed, air pressure is sent to the clutch release chamber, which pushes on the clutch release fork to disengage the clutch. If the pedal is depressed further the clutch brake valve will allow air pressure to operate the clutch brake. This system can be used with most standard transmissions as a separate system or altered to work in conjunction with an electronically controlled automated standard transmission.

B. Transmission Diagnosis and Repair (16 Questions)

Task B1   Determine the cause of transmission component failure, both before and during disassembly procedures.

Determining the cause of transmission failure before disassembly is generally harder than when it is apart. A discussion with the driver may not lead you to the correct conclusion, but it will at least inform you of the speed, operating conditions, and possible noises that may have been heard when the failure occurred. A road test and a check of the transmission fluid level and condition will also help in the failure diagnosis. When the transmission is apart, chipped gear teeth, worn bearings, and shaft damage can easily be identified. Signs of heat, spalling, or cracks in the components will aid you in the correct failure analysis.

Although bearing failures can be caused by a number of operational and assembly conditions, most bearing failures are due to dirt, lack of lubrication or improper lubricant. Gear failures can also be related to dirt and lubrication as well as insufficient clearance whether caused by bearings, shaft alignment or twisting, or improper timing during assembly. Improper handling of gears prior to assembly can also cause improper gear operation and reduced gear life or failure. Mainshaft failures due to twisting are usually caused by shock loading or overstressing the shaft by starting in too high of a gear. Input and output shaft failures are generally due to improper operating angles, such as misalignment between the engine and transmission or improper driveline angularity.

Task B2   Diagnose transmission vibration/noise, shifting, lockup, slipping/jumping out-of-gear, and overheating problems; determine needed repairs.

Technicians should road test the vehicle to determine if the driver's complaint of noise is actually in the transmission. Also, technicians should try to locate and eliminate noise by means other than transmission removal or overhaul. If the noise does seem to be in the transmission, try to break it down into classifications. If possible, determine what position the gearshift lever is in when the noise occurs. If the noise is evident in only one gear position, the cause of the noise is generally traceable to the gears or bearings involved in the selected gear. Noise is generally caused by a worn, pitted, chipped, or damaged gear or bearing.

Vibrations are usually developed in the driveline and rarely caused by the transmission, although the transmission can transmit these to the cab and driver. Overheating is generally associated with lubrication problems, but it can also be caused by improper gear clearances and bearing failures. Shifting problems in the front section such as hard shifting, slipping out of gear, jumping out of gear, and locking-up can be caused by a variety of different conditions, such as tight or worn shifter linkage, bent shift forks, worn or damaged gear clutching teeth, worn or twisted mainshaft splines, or faulty detent and interlock mechanisms. Auxiliary section shift problems are usually related to the air supply, valves, or shift cylinders, but these can also be caused by mechanical faults such as a worn or faulty synchronizer, twisted shafts, worn clutching teeth, bearing failures, or improper driving techniques.

Task B3   **Inspect, adjust, repair, or replace transmission remote shift linkages, cables, brackets, bushings, pivots, and levers.**

Manual adjustment of the automatic transmission manual gear range selector valve linkage is important. The shift tower detents must correspond exactly to those in the transmission. Failure to obtain proper detent in DRIVE, NEUTRAL, or REVERSE gears can adversely affect the supply of transmission oil at the forward or fourth (reverse) clutch. The resulting low-apply pressure can cause clutch slippage and decreased transmission life.

The effort required to move a manual transmission gear lever from one gear position to another varies. If too great an effort is required, it is a constant cause of complaint from the driver. Most complaints are with remote-type linkages used in cab-over-engine vehicles. Before checking for hard shifting, the remote linkages should be inspected. Linkage problems stem from worn connections or bushings, binding, or improper adjustment, lack of lubrication on the joints, or an obstruction that restricts free movement.

Task B4   **Inspect, test operation, adjust, repair, or replace air shift controls, lines, hoses, valves, regulators, filters, and cylinder assemblies.**

Main box shifts in most standard transmissions are mechanical operations while gear selection in the auxiliary section are air controlled. The air system requires a number of components such as; an air filter/pressure regulator assembly, slave valve, control valve (gearshift handle), range shift cylinder, splitter cylinder and air lines to supply and interconnect the shift components. The system operates relatively trouble free with proper air system maintenance. Moisture, oil, alcohol, and dirt circulation can cause many shift problems from slow or delayed shifts to no shifts as well as component failure.

Contaminants will affect the operation of the air filter/regulator assembly. Any shift problem diagnosis should begin with an air pressure check at the regulator outlet. Most transmissions regulate air pressure at approximately 60 psi. Although the driver initiates all shifts, the range shifts only take place when the transmission is in neutral and a slave valve directs air to the appropriate side of the range cylinder. When a failure occurs in the air supply to the range shift cylinder, the auxiliary section will remain in the previously selected range. The splitter cylinder receives a constant supply of air to the front side of the piston, holding the yoke bar rearward. When a control valve or air line failure occurs, the splitter cylinder will either remain static or shift rearward and stay in this position until the failure is repaired. O-ring failures on the cylinder pistons will usually prevent or slow shifts. O-ring failures on the yoke bars will not only affect the shift but also allow pressurized air into the transmission case, which may cause exterior oil leaks if the breather is clogged. When replacing components, follow the manufacturer's air line routing schematics to ensure proper system operation.

Task B5   **Inspect, test operation, adjust, repair, or replace electronic shift controls; shift, range and splitter solenoids; shift motors; indicators; speed and range sensors, electronic/transmission control units (ECU/TCU); neutral/in-gear and reverse switches, and wiring harnesses.**

An electric shifter assembly is used to replace the manual shifter lever. The transmission shift assembly or shift finger works in a typical shift rail housing. This assembly includes a shift finger that is automated by two reversing DC electric motors. One motor controls rail selection and the other controls shift collar movement. The motors rotate ball screw assemblies to move the finger back and forth for gear selection or side to side to one of the three shift rails. Rail select and gear select sensors are used to communicate the selected gear to the electronic control unit (ECU). An electronically controlled, air operated range valve is used to replace the driver controlled range valve. The computer energizes a pair of solenoids (one high range and one low range) to control the airflow to the range valve. When these solenoids are de-energized air is exhausted and the range cylinder will remain in the previous selected position. An electronically controlled, air operated splitter valve replaces the driver controlled splitter valve. Speed sensors are used to signal input shaft, mainshaft, and output shaft speed to the computer. All three sensors are inductive pulse style. These use a magnetic pick-up and a tone wheel to generate frequency and voltage that the computer can translate as shaft speed.

Electronically controlled transmissions that are experiencing problems require road testing and the use of electronic service tools (EST) to test the operation of the various components. Wiring harnesses and connectors require a careful inspection to eliminate the chance of component replacement due to a wiring problem. Always follow the manufacturer's recommended procedures for adjustment, repair, or replacement of any electronic components. Many electronic components are susceptible to static electricity damage; grounding your body to the vehicle may be required.

Task B6   Inspect, test operation, repair, or replace electronic shift selectors (in-cab controls), air and electrical switches, displays and indicators, wiring harnesses, and air lines.

Once wiring harnesses and electronic control units have been checked and their integrity is not in question, electronic shift selectors, displays, and indicators can be identified as faulty or out of adjustment. With the exception of some shift selector adjustments most components require replacement. Check the air supply to air switches, and ensure that no contamination is present at the switch. Contamination or insufficient air supply to an air switch will affect its operation. Air line replacement or cleaning could prevent the possibility of replacing a good component.

Task B7   Use appropriate diagnostic tools and software, procedures, and service information/flow charts to diagnose automated mechanical transmission problems; check and record diagnostic codes, clear codes, interpret digital multimeter (DMM) readings, determine needed repairs.

A variety of tools can be used to diagnose electronically controlled transmissions. The system's self-diagnostic blink codes, digital multimeters, and electronic service tools (hand-held scanners, laptop computers, etc.) can be used to retrieve fault codes and check circuit integrity. Some fault codes can be set by intermittent voltage irregularities or the component operating beyond its pre-set parameters. These codes may be erased, but the condition that set the code should be identified. Some codes are hard and cannot be erased until the problem component is repaired. All readings and codes should be recorded and compared to the manufacturer's specifications. Back probing connectors can be performed with proper meter leads, which eliminates the need to probe through the wire insulation. Analog meters and test lights should not be used when working with electronic components and circuits. Only use these if they are recommended.

All measurements and procedures should follow the manufacturer's diagnostic flow charts to ensure that the proper fault diagnosis is obtained and no electronic components are damaged due to improper test procedures.

Task B8   Diagnose automated mechanical transmission problems caused by data link/bus interfaces with related electronic control systems.

Electronically automated mechanical transmissions use data links/interfaces to connect the transmission to the various vehicle systems that must work along with it. With many components and wiring SAE J 1939 compliant, the transmission electronics can communicate operational data as well as faults with these other systems. The fault codes are logged both in the transmission manufacturer and SAE formats and can usually be read through on-board diagnostic service lights or readouts or by the use of handheld diagnostic electronic service tools.

Task B9   Remove and replace transmission; inspect and replace transmission mounts, insulators, and mounting bolts.

Transmission mounts and insulators play an important role in keeping drivetrain vibration from transferring to the chassis of the vehicle. If the vibration were allowed to transmit to the chassis of the vehicle, the life of the vehicle would be greatly reduced. Driving comfort is another reason why insulators are used in transmissions. The most important reason is the ability of the insulators to absorb shock and torque. If the transmission was mounted directly to a stiff and rigid frame, the entire torque associated with hauling heavy loads would need to be absorbed by the transmission and its internal components, causing increased damage and a much shorter service life. Broken transmission mounts are not readily identifiable by any specific symptoms. They should be visually inspected for missing mount bolts, swelling and cracks in the rubber. To check a mount assembly that is not visibly damaged or worn, apply the parking brakes, start the engine and place the transmission in low gear to check the left side mount and reverse gear to check the right side. Partially engage the clutch to place a load on the driveline. This driveline torque should produce a lifting force at each mount. Worn or broken transmission mounts will allow visible movement at the mount assembly.

Task B10   Inspect for leakage and replace transmission cover plates, gaskets, sealants, seals, vents, and cap bolts; inspect seal surfaces.

In diagnosing and correcting fluid leaks, finding the exact cause of the leakage can be difficult because evidence of the leakage may occur in an area other than the source of the leakage. To assist in locating a leak, thoroughly wash the transmission and add leak detection dye to the transmission oil. Road test the vehicle to allow for dye circulation throughout the transmission. Use an ultraviolet/black light to inspect the transmission for leaks. If a leak is present, the source should be evident because any seeping dyed lubricant should glow. Do not replace transmission gaskets with sealant. Gaskets located between housings can provide operational clearance for components as well as sealing. In automatic transmissions silicone can be drawn into the hydraulic system and block circuits and pump screens. Always check the transmission breather filters when repairing any transmission leaks. Under normal operating conditions the transmission can build up pressure inside the case if the vents or breathers are not functioning correctly. This pressure can cause fluid to leak past seals that do not need replacement or seals that are otherwise in good working order or cause damage to seals and gaskets causing leaks.

Task B11   Check transmission fluid level, and condition; determine needed service, and add proper type of lubricant.

Most manufacturers suggest a specific grade and type of transmission oil, heavy-duty engine oil, or straight mineral oil, depending on the ambient air temperature during operation. Do not use mild EP gear oil or multipurpose gear oil when operating temperatures are above 230° F (110° C). Many of these gear oils break down above 230° F (110° C) and coat seals, bearings, and gear with deposits that might cause premature failures. If these deposits are observed (especially on seal areas where they can cause oil leakage), change to heavy-duty engine oil or mineral gear oil to assure maximum component life.

Always follow the manufacturer's exact hydraulic fluid specifications. For example, several transmission manufacturers recommend DEXRON, DEXRON II, and type C-3 (ATD approved SAE 10W or SAE 30) oils for their automatic transmissions. Type C-3 fluids are the only fluids usually approved for use in off-highway applications. Type C-3 SAE 30 is specified for all applications where the ambient temperature is consistently above 86° F (30° C). Some, but not all, DEXRON II fluids also qualify as type C-3 fluids. If type C-3 fluids must be used, be sure all materials used in tubes, hoses, external filters, seals, etc., are C-3 compatible.

Transmission manufacturers are recommending the use of synthetic oils in their current models because of their improved lubrication qualities and longer lifespan. If synthetic oil is used, most warranty periods are increased substantially.

Changing the transmission fluid is a valuable maintenance procedure for all transmissions that should not be neglected. The leading cause of standard transmission bearing failure and wear of shafts and gears is the circulation of dirty oil. When using oil that promotes extended oil change intervals, the oil should be inspected for dirt whenever the fluid level is checked. Automatic transmissions should have the oil level checked daily and the color, smell and signs of particles should be observed. Slipping hydraulic clutches will wear clutch discs and overheat the fluid producing a burnt smell and a blackened color from the friction material. Overfilling both standard and automatic transmissions can produce leaks, overheating and component wear. Moving parts striking and whipping the oil can cause aeration. Aerated oil does not lubricate and cool as efficiently as pure oil can so friction and temperature will increase. The hot, thinner oil and increased oil movement in the housing can cause oil leaks at vents and shaft seals.

Task B12   Inspect, adjust, and replace transmission shift lever, cover, rails, forks, levers, bushings, sleeves, detents, interlocks, springs, and lock bolts/safety wires.

When a sliding clutch is moved to engage with a main shaft gear, the mating teeth must be parallel. Tapered or worn clutching teeth try to "walk" apart as the gears rotate, causing the sliding clutch and gear to slip out of engagement. Slipout generally occurs when pulling with full power or decelerating with the load pushing. Different from slipout, jumpout occurs when a fully engaged gear and sliding clutch are forced out of engagement. It generally occurs when a force sufficient to overcome the detent spring pressure is applied to the yoke bar, moving the sliding clutch to a neutral position. Keep in mind that the whipping action of extra long or heavy shift levers can cause the transmission to jump out of gear.

The shift detent mechanism consists of spring-loaded balls that are located in the shift cover and rest in notches in the shift rails. Each shift rail has three notches, one for each gear selection position. The gearshift must overcome the tension of the detent ball and spring to force the ball upward and allow the rail to move. If the ball becomes worn or gouged, the notches become worn or the detent spring breaks the detent can fail to operate correctly and the shift rails resistance to move will decrease allowing vibrations and road shock to disengage the clutch collar from the gear (jumping out of gear). The detent springs must be installed before installing the gearshift and they should be installed dry.

Task B13   Inspect and replace input shaft, gears, spacers, bearings, retainers, and slingers.

The input shaft can be affected by many different parts of a drivetrain. On a vehicle with a manual transmission, the input shaft fits into the pilot bearing, splines into the clutch disc or discs, providing a place for the clutch release bearing to move along. Whether manual or automatic transmission, this vital part of the drivetrain should be inspected for any abnormal wear of the splines as well as the gear portion of the input shaft. This is one of the few parts of the drivetrain that carries the entire torque load of the vehicle.

Input shafts are susceptible to damage from shock loading and vibrations due to incorrect driveline angularity. These conditions can produce spline wear and cracks, drive tooth damage and twisted or broken shafts. Bell housing misalignment can also lead to premature spline wear and pilot and input bearing failure.

Task B14   **Inspect main shaft, gears, sliding clutches, washers, spacers, bushings, bearings, auxiliary drive gear/assembly, retainers/snap rings, and keys; determine needed repairs.**

Washers, spacers, bushings, and bearings rely on good lubrication during operation. These components should be inspected for scoring and discoloration due to lack of lubrication and end thrust. Bearings and bushings require inspection for pitting caused by dirt, flaking or spalling of the bearing surface caused by fatigue and damage from improper installation. Vibration can cause fretting of the outer bearing race. This occurs as the bearing bore pattern is transferred to the bearing which leaves the appearance of slight scoring or lines. Snap rings and retainers can lose tension or be damaged during removal and installation. If the snap rings are distorted or do not fit securely in place, they must be replaced. Mainshafts should be inspected for spline wear from sliding clutch operation and twists from shock loading which can cause hard or no shifting of the sliding clutches. The sliding clutches inner splines should be checked for wear and the engaging teeth for damage from partial engagement. Gear damage can be caused by a variety of conditions. Dirt and filings circulated in the oil can cause pitting of the gear teeth. Compare the amount of pitting to wear charts before condemning the gear. Improper installation or removal procedures and shock loading can cause cracks in gears. Check gear clutching teeth for shortness, taper, and damage to their beveled edges which could cause incomplete engagement or jumping out of gear. Inspect gear teeth for any bumps or swells caused by improper handling or installation.

Two other types of gear damage are bottoming and climbing. Bottoming occurs when the teeth of one gear touch the lowest point between the teeth of a mating gear. Bottoming does not occur in a two-gear drive combination but can occur in multiple-gear drive combinations. A simple two-gear drive combination always tends to force the two gears apart; therefore, bottoming cannot occur in this arrangement. Climbing is caused by excessive wear in gears, bearings, and shafts. It occurs when the gears move sufficiently apart to cause the apex (or point) of teeth on one gear to climb over the apex of the teeth on another gear with which it is meshed. This results in a loss of drive until other teeth are engaged, and causes rapid destruction of the gears.

Task B15   **Inspect countershafts, gears, bearings, retainers/snap rings, and keys; adjust bearing preload/end play; time multiple countershaft gears; determine needed repairs.**

All twin countershaft transmissions are "timed" at assembly. It is important that the manufacturer's timing procedures are followed when reassembling the transmission. Timing assures that the countershaft gears contact the mating main shaft gears at the same time, allowing main shaft gears to center on the main shaft and equally divide the load. Timing is the simple procedure of marking the appropriate teeth of a gear set prior to removal (while they are still in the transmission). In the front section, it is necessary to time only the drive gear set. Depending on the model, the low range, deep reduction, or splitter gear set is timed in the auxiliary section. Component inspection for the countershaft gears, retainers/snap rings, and keys should be followed as described in the previous task.

Task B16   **Inspect output shaft, gears, washers, spacers, bearings, retainers/snap rings, and keys; determine needed repairs.**

Transmission output shaft inspection includes checking all splines for wear and twists. The output shaft is the first transmission component to receive vibration and operational stress from the drive train, which can result in, wear on its bearing surfaces, washers, spacers and bearing retainers. Bearing and thrust washer wear can allow axial movement of the output shaft during operation. This axial movement is a common cause of rear transmission seal leaks and auxiliary case damage. The specification for most output shaft endplay is in the range of 0.005 inch to 0.012 inch. Examine the range gear bearing surface(s) on the output shaft for any roughness and scoring. Inspect washers, spacers, bushings, and bearings for scoring and discoloration due to lack of lubrication and end thrust, pitting caused by dirt, flaking or spalling of the bearing surface by fatigue, and damage from improper installation. All components should be compared to the transmission manufacturer's wear charts and diagnosis manuals for component comparisons and diagnosis verification.

Task B17   **Inspect reverse idler shaft(s), gear(s), bushings, bearings, thrust washers, and retainers/snap rings; check reverse idler gear end play; determine needed repairs.**

Most of the gears in a twin countershaft transmissions are either floating or pressed on their shafts, which does not generate much if any shaft wear. The reverse idler gear is bearing mounted. Since it is not located directly between two shafts, separating forces will be developed when under load. Constant bearing operation and the separating forces generated during reverse gear operation can produce bearing wear and wear on the idler shaft where the bearing rides. The bearing and shaft condition will depend on the amount of lubrication the assembly received during its lifespan and the condition and type of lubricant used. The thrust washers and retainers should be inspected for wear, as these components will have an effect on the amount of idler gear endplay.

Task B18   **Inspect synchronizer hub, sleeve, keys/inserts, springs, blocking rings, synchronizer plates, blocker pins, and sliding clutches; determine needed repairs.**

Check the synchronizer for burrs, uneven and excessive wear at contact surfaces, and metal particles. Check the blocker pins for excessive wear or looseness. Check the synchronizer contact surfaces for excessive wear. If the vehicle is equipped with cone-type synchronizers, check to see that the blocker ring is within tolerance by twisting the ring onto the matching gear cone. If the blocker ring "locks" itself onto the gear surface, the ring is still useable.

Task B19   **Inspect transmission cases including mating surfaces, bores, bushings, pins, studs, vents, and magnetic plugs; determine needed repairs.**

Two of the more simple items on a transmission that often get overlooked during servicing are the transmission case and breather(s). The transmission case must be checked for any signs of fatigue. Fatigue and housing misalignment are causes of cracks in the main box housing. Cracks in the auxiliary housing are generally due to vibrations and stress from improper driveline angularity. Cracking is a symptom that is usually accompanied by fluid leakage. Plugged breathers are also associated with fluid leakage but not at the location of the breather. When a breather becomes plugged, fluid is often forced past seals in the transmission. If a technician jumps to a conclusion when they notice fluid leakage at a seal, they may mistakenly replace the seal and think the problem is solved. A thorough job requires the technician to check the transmission breathers during any transmission diagnosis. To clean a breather it should be soaked in solvent to soften and dilute the blockage and afterwards blown out with compressed air to remove any remaining debris.

Task B20   **Inspect, service, or replace transmission lubrication system components, pumps, troughs, collectors, slingers, coolers, filters, and lines and hoses.**

Many standard transmissions rely on splash lubrication from the rotation of the gears. The lower gears that contact the oil act like paddle wheels that pick up oil and transfer and splash it on the upper gears. Some transmissions use troughs to catch splashed oil and direct it to critical wear areas that may not receive much splash. This lubrication system requires little service and inspection other than ensuring all troughs are clean, not damaged, securely in place, and that the oil is changed regularly, the proper oil is used, and the correct fluid level is maintained. Other standard transmissions use a combination of splash lubrication and an oil pump to circulate oil to bearings and areas of high wear. The oil pump can be located internally or externally and may use an external oil cooler. With these systems the pump should be disassembled an inspected for gear and housing wear and the proper operating clearances. The oil cooler should be flushed and checked for leaks. All standard transmissions use slingers to limit the amount of oil present at the seal areas. Oil leaks can occur when these are not located correctly, bent, or damaged.

In automatic transmissions, the oil pump draws fluid through a sump filter/screen and circulates the fluid through the torque converter, oil cooler, lubrication circuits, and to various valves and clutches to obtain hydraulic gear selection and correct fluid temperature. The pumps must be inspected as described above and the torque converter, all passageways and hoses must be flushed thoroughly. The sump filter should be changed during rebuild and at recommended service intervals. To service the oil pump the transmission and torque converter must be removed before the pump can be removed.

Task B21   Inspect, test, replace, and adjust speedometer drive components (mechanical and electronic).

Mechanical speedometer drives consist of a spiral drive gear located on the output shaft, a speedometer driveshaft/pin and an adapter that drives the speedometer cable. The adapter must be selected for the final drive ratio and the tire circumference to ensure the proper speedometer reading is obtained. To check the operation of the speedometer drive, block the wheels to prevent the truck from moving, raise one drive wheel, remove the speedometer cable, release the park brakes and have another technician rotate the drive wheel while observing the adapter drive pin. The pin should rotate as the wheel is turned. If the pin is stationary, remove the adapter and ensure that the drive pin in the transmission is rotating. If this pin is rotating, the adapter is faulty. If the pin does not rotate, it must be removed and replaced due to sheared teeth on its drive gear.

Electronic speedometer drives use a pulse generator consisting of a 16 tooth pulse wheel and a magnetic sensor. As the output shaft turns, the teeth cut the magnetic field of the coil, inducing AC voltage pulses. The higher the output shaft speed the higher the AC voltage. The sensor can be tested by measuring its coil resistance with an ohmmeter. The operation of the sensor can be checked with electronic service tools or with an AC voltmeter and manually rotating the drive wheel as described above. A damaged pulse wheel can also cause improper sensor operation.

The best way to determine whether or not a problem in any system is due to electrical or mechanical failure is to gather information about the system. Considerations like whether the power supply to that system is shared with another system, or whether grounding to the component is shared, are both quick ways to help pinpoint the source of the fault. Mechanical failure in a system is almost never intermittent and usually can be investigated by a simple visual inspection.

Task B22   Inspect, adjust, service, repair, or replace power take-off assemblies and controls.

Power take-off (PTO) units are designed to drive other components such as hydraulic pumps and driveshafts to drive various devices. Inspection of the PTO can be done by visually checking it for leaks and damage and listening for bearing or gear noise. To thoroughly inspect a PTO, it requires removal of the unit and carefully inspecting it. The most common problem is with adjustment or binding of the control linkage, cables, or levers. When removing and installing transmission mounted units, shims may be required to obtain the correct gear mesh with the transmission PTO drive gear.

Transmission driven PTO pumps are usually operated in neutral so pump speed can be controlled. Operation in gear could produce erratic PTO operation and shifting difficulty due to the load of the pump on the countershaft. Incorrect PTO driveshaft angularity, joint wear and incorrect shaft phasing can generate vibrations during operation.

Task B23 Inspect and test function of backup light, neutral start, and warning device circuit switches.

Mechanical standard transmissions generally use a spring loaded normally open back-up light switch that is closed by the shift rail lifting the ball which closes the switch when reverse is selected. This switch can also be used to control backup warning devices. To test this switch, remove the electrical connector and with an ohmmeter take switch readings with the selector in neutral and in reverse. The reading should be infinite in neutral and 0 ohms in reverse. Automatic transmissions use multi-position neutral start and reverse gear switches on the linkage. These switches are open in all forward gear positions but close the start circuit in the park and neutral position and close the backup light circuit in the reverse position. Electronically automated standard transmissions and electronically controlled automatic transmissions use gear position sensors or switches to signal the electronic control unit of the transmission gear status. Electronic service tools are used to verify the operation of these by sensors and switches. Manufacturer's diagnostic charts should be followed for correct test and diagnosis procedures.

Task B24 Inspect and test transmission temperature sending unit/sensor; determine needed repairs.

The most reliable way to test a transmission temperature sensor for accuracy is to obtain a temperature-to-resistance chart from the manufacturer. This allows the technician to determine if the temperature sensor is sending a signal appropriate for the temperature it is encountering. Another way of diagnosing the temperature indicating circuit is to place a variable resistance in the place of the sensor. Varying a known resistance and checking the temperature display for correspondence is also a good way of helping pinpoint the source of the problem.

If a temperature sensor is computer controlled, this type of sensor usually receives a reference voltage of 5 volts from the management computer and return a portion of this voltage as a signal that varies depending on the temperature it is operating at. These sensors should be checked for the input voltage and the output signal voltage with a digital multimeter or an electronic service tool.

High temperature gauge readings can be correct if the vehicle is operating under extreme conditions for an extended period of time. A rise in temperature readings for a short period of time after the vehicle has shut down is normal because oil circulation and air flow around the transmission has stopped.

Task B25 Inspect, adjust, repair, or replace transfer case assemblies and controls.

A transfer case is simply an additional gearbox located between the main transmission and the rear axle and may be of single or two speed design. Two shift forks are commonly used, one for selection between low, neutral, and high range and the other for selecting and de-selecting drive to the front axle (front axle de-clutch). The drive to the rear axle is constant except when neutral is selected. Neutral position is usually selected for PTO operation if equipped. Drive to the front axle is only available when selected.

The transfer case may be equipped with an optional parking brake, PTO, and a speedometer drive gear that can be installed on the idler assembly. Most transfer cases that use the countershaft design in their gearing are of the constant mesh helical cut type. Most countershafts are mounted on ball or roller bearings. All rotating and contact components of the transfer case are lubricated by oil from gear throw-off during operation. However, some units are provided with an auxiliary oil pump, externally mounted to the transfer case. To diagnose components in the transfer case, use the same logic as drive axle or transmission gearing.

Task B26 Diagnose internal and external transmission inertia brake problems; determine needed repairs.

The function of the inertia brake is to stop the countershaft of an electronically automated standard transmission for initial gear engagement and to slow the countershaft during up-shifts. The inertia brake uses an electronically controlled air solenoid to control the operation of the brake. When diagnosing an inertia brake failure, the electronic control system, air system, and the mechanical brake must all be considered.

C. Driveshaft and Universal Joint Diagnosis and Repair (9 Questions)

Task C1   Diagnose driveshaft and universal joint noise, vibration, and runout problems; determine cause of failure and needed repairs.

Often vibration is too quickly attributed to the driveshaft. Before condemning the driveshaft as the cause of vibration, the vehicle should be thoroughly road tested to isolate the vibration cause. To assist in finding the source, ask the operator to determine what, where, and when the vibration is encountered. It is very helpful to keep in mind some of the causes of driveline vibration. U-joints are the most common source if the vibration is coming from the driveshaft, while driveshaft balancing is the next most common. Pay special attention to phasing when removing or installing a driveshaft.

Driveshaft phasing is the positioning of the tube yoke and slip yoke in-line with each other. Phasing a driveshaft times its speed fluctuations to input the final drive at the correct time to obtain a constant drive pinion speed. Always check with the manufacturer's manuals to be sure of the proper yoke positioning. The driveshaft should also be checked for run out, which is a bend in the tube. Speed fluctuations plus any shaft run out will develop inertia at higher speeds. This inertia will create vibrations that can be destructive to the driveshaft and its drive and driven components. When inspecting the driveshaft on the truck, rotate the shaft with a dial indicator against the tube to identify any run out. Check the specifications for the proper measuring locations and the allowable runout limits of the shaft. The yokes should also be checked for run out with a dial indicator and vertical alignment with a protractor or inclinometer. The yokes should fit their shafts firmly without free play and be aligned with their shafts. Use the manufacturer's procedures and specifications for these tests. Don't mistake output shaft or pinion shaft radial play for yoke play.

Task C2   Inspect, service, or replace driveshaft, slip joints/yokes, yokes, drive flanges, universal joints, driveshaft boots and seals, and retaining hardware; properly phase/time yokes.

The following are descriptions of common visually evident damage to U-joints. Cracking shows up as stress lines due to metal fatigue. Galling occurs when metal is cropped off or displaced due to friction between surfaces, most commonly found on trunnion ends. Spalling occurs when chips, scales, or flakes of metal break off due to fatigue rather than wear. Spalling is usually found on splines and U-joint bearings. Pitting appears as small pits or craters in metal surfaces due to corrosion. Pitting can lead to surface wear and eventual failure. Brinelling is evident by grooves worn in the bearing race surface, often caused by improper installation of the U-joints. Do not confuse the polishing of a surface where no structural damage occurs with actual brinelling.

When replacing a universal joint, a universal joint puller is recommended but many technicians use a jack to press the universal joints from the trucks yokes and hammer the joints from the shaft in a vice. Care must be taken to ensure that no damage is incurred when jacking or hammering on the driveshaft yokes. When the joint is removed clean the bearing cup bore and file of any nicks that may affect the proper installation of the cups. When installing, always start the bearing cups on the trunnions by hand to prevent damaging the bearings inside the cups. When lubricating universal joints, use a lithium based, extreme pressure grease meeting NLGI grade 1 or 2 specifications. Grease should be pumped into the joint until clean grease appears at all four trunnion seals. Before removing a slip joint from a driveshaft, always mark both pieces to allow for replacement in the same position so shaft balance is not affected. Inspect the splines on both the shaft and in the slip joint for cracks, damage, and wear. During reassembly position the tube yoke and slip yoke in-line with each other to properly phase the shaft. When lubricating the slip yoke, use an extreme pressure grease meeting NLGI grade 1 or 2 specifications or U-joint grease can be used because it should exceed these specifications. The grease should be applied until fresh grease is present at both the slip yoke seal and the pressure relief hole.

Task C3 Inspect, and replace driveshaft center support bearings and mounts.

Center bearings are lubricated by the manufacturer and are not serviceable. However, when replacing a support bearing assembly, be sure to fill the entire cavity around the bearing with waterproof grease to shield the bearing from water and contaminants. Grease must fill the cavity to the extreme edge of the slinger surrounding the bearing. Use only waterproof lubricants after consulting a grease supplier or the bearing manufacturer for recommendations. Also pay attention when removing the existing center support bearing. Any shims used to adjust driveline angle must be reinstalled when installing a replacement bearing.

Task C4 Measure and adjust vehicle ride height; measure and adjust driveline slopes and angles (vehicle loaded and unloaded), including PTO driveshafts.

With the vehicle on a level surface, tire pressures equalized, and the output yoke on the transmission in a vertical position, use either a magnetic base protractor or an electronic inclinometer to measure driveshaft angle. Always measure driveshaft angle from front to rear. Always take driveline angle measurements with the vehicle at the correct ride height loaded and unloaded.

To correct universal joint operating angles the angle of the transmission and/or final drive(s) must be changed. Incorrect transmission angle is usually due to worn engine and transmission mounts. New mounts should be installed to restore the correct angle. To adjust the final drive angle, shims can be added or removed from the torque rods to rotate the axle pinion to the correct angle. On leaf spring suspensions, shims between the axle and the leaf spring can be added or removed to adjust the pinion angle. Maximum operating angles depend on shaft operating rpm. Refer to angle charts or manufacturer's angularity software for the maximum and recommended operating angle. The difference between the operating angles at each end of a driveshaft should be less than 1 degree to minimize vibration. Before measuring driveline angles, inspect all suspension components and mounts for wear and looseness. Loose U-bolts and worn bushings can allow movement during operation while allowing for correct angle measurements when stationary.

Task C5 Use appropriate driveline analysis software to diagnose driveline problems.

A vehicle's driveline can be the source of noise, vibration, or running gear damage. If the driveline angularity is not correct, torsional accelerations and inertia can create these symptoms. The angularity can be checked and calculated manually or with a computer software program. These software programs are valuable tools that provide a quick and accurate means of diagnosing and correcting improper driveline angularity.

Task C6   Diagnose driveline retarder problems; determine needed repairs.

Driveline retarders are a common type of auxiliary braking device. Hydraulic retarders are commonly used on vehicles equipped with automatic transmissions. These units produce very little wear due to the fact that there is no contact between the moving rotor or paddle wheel and the stationary vaned housing. They do however suffer from failures related to the heat that is developed during retarder operation. Hydraulic retarders utilize the transmissions heat exchanger although some vehicles may require additional coolers. Other operational problems can be caused by sticking or failed valves, leaking seals or the air supply system. A thorough knowledge of the retarder and its controls are the key to correct diagnosis.

D. Drive Axle Diagnosis and Repair (11 Questions)

Task D1   Diagnose drive axle unit noise and overheating problems; determine needed repairs.

Because of the similar nature of transmissions, drive axles, and transfer cases, determining exactly where a noise is coming from and which component is causing the noise may be very difficult. A set of guidelines may work very well for certain combinations of components, while not working at all for others. Sometimes it may be beneficial to keep a list describing sounds that the truck exhibited and the component that caused it. This list could prove to be a good way to assist a technician's memory when a vehicle is currently in for repair and the cause is not evident. One rule that does apply to all situations: technicians must always bear in mind that universal joints, transmissions, tires, and drivelines can create noises that often sound alike.

Drive axle temperatures can be adversely affected by operating under heavy loads and high speeds, incorrect adjustment or failure of any of the axle assembly bearings or gears, incorrect type or poor quality axle fluid, low fluid levels or contaminated fluid. All of these conditions will increase the amount of friction and heat generated in the final drive.

Task D2   Check and repair fluid leaks; inspect and replace drive axle housing cover plates, gaskets, sealants, vents, magnetic plugs, and seals.

A technician can usually determine the operating condition of the drive axle differential by the fluid. Pay special attention to the condition of the fluid during the scheduled fluid changes. Most drive axles are equipped with a magnetic plug that is designed to attract any metal particles suspended in the gear oil. A nominal amount of "glitter" is normal because of the high torque environment of the drive axle. When a small amount of glitter is found, the customer should be informed of this and asked to continue to monitor the amount. Too much "glitter" indicates a problem that requires further investigation. Carefully investigate any source of leaks found on the axle differential. Replace the seal if the axle differential breather is plugged does not cure leaking seals. A plugged breather results in high pressure, which can result in leakage past seals.

Task D3   Check drive axle fluid level, and condition; determine needed service; add proper type of lubricant using correct fill procedure.

Drain and flush the factory-fill axle lubricant of a new or reconditioned axle after the first 1,000 miles (1,609 km) and never later than 3,000 miles (4,827 km). This is necessary to remove fine particles of wear material generated during break-in that would cause accelerated wear on gears and bearings if not removed. Draining the lubricant while the unit is still warm ensures that any contaminants are still suspended in the lubricant. Flush the axle with clean axle lubricant of the same viscosity as used in service. Any time that the fluid is found to be a grayish or whitish milky appearance this is usually caused by moisture contamination. This is generally a condensation problem found in trucks that are seldom driven and only for short trips. The condensation cannot boil off and accumulates in the axle. When this occurs, the fluid must be drained and the housing flushed. The wheels should also be removed and the bearings and cavity cleaned before filling the axle with fresh lubricant. The fluid level is correct when the fluid is level with the bottom of the oil fill port and the hubs have been refilled. Do not flush axles with solvents such as kerosene. Avoid mixing lubricants of a different viscosity or oils made by different manufacturers.

Task D4   Remove and replace differential carrier assembly.

When removing the differential carrier, leave two of the mounting bolts in the axle housing to hold the differential in place until a hydraulic jack can be placed under the differential carrier. The extreme weight of the differential carrier must be safely strapped onto the hydraulic jack for removal of the carrier. When the carrier is ready for installation, inspect the mounting surfaces for nicks, scratches, or gouges.

Task D5   Inspect and replace differential case assembly including spider/pinion gears, cross shaft, side gears, thrust washers, case halves, and bearings.

There are many types of damage that can occur in a differential assembly, including shock failure. This damage occurs when the gear teeth or pinion are stressed beyond the strength of the material from which they are machined. The failure may be immediate from a sudden shock or it may be a progressive failure after the initial shock cracks the teeth or pinion. As in any other situation, early detection of damage can prevent additional damage.

Noises that occur when cornering are usually related to the differential gearing. The spider/pinion gear and side gear thrust washers are prone to extreme wear when axle spin-out occurs. The separating forces of the gear and the high speed rotation of the gears inside the case create tremendous heat at their thrust washers, which removes the oil, and scoring of the gears, case, and washers occur. The rotation of the differential gearing can also produce wear on the cross shaft. Differential case bearings are susceptible to damage from filings, dirt, and moisture contaminated axle lubricant. Bearings should be inspected for pitting, fatigue, lines, and scoring.

Task D6   Inspect and replace components of locking differential case assembly.

The typical single reduction carrier with differential lock has the same type of gears and bearings as the standard type carriers. However, an air-actuated shift assembly that is mounted on the carrier and operated from the truck's cab operates the differential lock. By actuating an air plunger or electric switch usually mounted on the instrument panel, the driver or operator can lock the differential to achieve positive traction and control of the truck on poor or slippery road or highway conditions.

To lock the differential, a shift collar is moved into mesh with splines on the differential case. This locks the axle shaft to the case, which prevents any movement of the differential gears, providing drive to both axle half shafts. The air supply, and shift assembly should be inspected for proper operation and the shift fork should be checked for wear and bends. Since the shift assembly is air engaged and spring released any air leaks at the piston or into the axle housing during lock operation could allow partial lock collar engagement and possible disengagement under heavy loads. Failure to release can be caused by a broken return spring or a seized apply piston. The shift collar, axle shaft, and case splines should be checked for wear and damage.

Task D7   Inspect differential carrier/case and caps, side bearing bores, and pilot (spigot, pocket) bearing bore; determine needed repairs.

Differential bearing caps should be marked before removal to ensure that they are reassembled onto the proper leg. Put the differential bearing outer race and adjusting nut in place before installing the bearing cap onto the leg of the differential. Placing the bearing race and adjusting nut in position before the bearing cap assures proper alignment of the bearing cap. This also makes the process of setting the differential bearing preload easier because the adjusting nut will move more freely inside of the leg and bearing cap.

Task D8   Measure ring gear run out; determine needed repairs.

The ring gear should rotate true. Run out is the amount of wobble or distortion in the ring gear. There are a number of different causes of run out some of which are: improper installation of the ring gear on the differential case, case damage on the ring gear mounting surface, and from operating under extreme loads. Run out can be checked by mounting a dial indicator on the carrier mounting flange and placing the pointer on the back of the ring gear. With the indicator set at zero, rotate the ring gear while noting the indicator readings. The maximum reading is the amount of ring gear run out. Compare the reading obtained to the manufacturer's specification. Typical runout limits are 0.006 inch to 0.008 inch. To correct run out, remove the case from the carrier and the ring gear from the case. Inspect all final drive components (not only the case) for damage or debris that could cause the run out. If the run out was caused by installation conditions the ring gear can usually be reinstalled. Ring gears distorted by loads are generally not reusable.

Task D9   Inspect and replace ring and drive pinion gears, spacers, shims, sleeves, bearing cages, and bearings.

Correct tooth contact between the pinion and the ring gear cannot be over-emphasized because improper tooth contact can lead to early failure of the axle, and noisy operation. Used gearing usually does not display the square, even contact pattern found in new gear sets. The gear normally has a pocket at the toe-end of the gear tooth that falls into a contact line along the root of the tooth. The more a gear is used, the more the line becomes the dominant characteristic of the pattern. If a ring and pinion is to be reused, measure the tooth contact pattern and backlash before disassembling the differential.

When removing a ring gear from a differential case, rivets should not be removed with a hammer and chisel because this can enlarge the rivet holes and damage the mounting flange. To prevent damage, rivets should be drilled and punched out. The ring gear can now be pressed off the differential case. On assembly, the ring gear must not be pressed onto the case because damage to the case and excessive run out can occur. The ring gear should be heated in water for expansion, then placed onto the case and rotated into alignment with the mounting holes. Use the original type of fastener and follow the manufacturer's recommended torque or installation procedures.

Task D10   Measure and adjust drive pinion bearing preload.

There are a few different methods of measuring pinion bearing preload. Most of the methods use a spring scale to measure the rolling resistance of the assembly under pressure. It is always best to refer to applicable manufacturer's specific information whenever testing for pinion bearing preload. When using the rolling resistance measurement method, note the spacer side used once the correct bearing preload has been established. Select a spacer 0.001 inch (0.254 mm) or larger for use in the final pinion bearing cage assembly procedures. The larger spacer compensates for slight growth in the bearings, which occurs when they are pressed on the pinion shank.

Task D11   Adjust drive pinion depth.

Before disassembling the differential, make drawings of the gear tooth contact pattern to ensure proper reassembly. A correct pattern is clear of the toe and centers evenly along the face width between the top land and the root. In other cases, the length and shape of the pattern are highly variable and are considered acceptable as long as the pattern does not run off the tooth at any time. If necessary, adjust the contact pattern by moving the ring gear and drive pinion. Ring gear position controls contact pattern along the face width of the gear tooth. Pinion position is determined by the size of the pinion bearing cage shim pack, and controls contact on the tooth depth of the gear tooth.

Task D12   Measure and adjust differential case/side bearing preload; set ring and pinion backlash.

There are a few different ways to measure and adjust differential bearing preload. Most of the methods use the same logic, although they go about it in different fashions. The idea is to place a dial indicator on the back face of the ring gear and adjust the differential bearing adjusting rings until there is no run out measured on the ring gear. When there is no run out measured on the ring gear, tighten the adjusting rings one or two notches further. This adds the proper amount of pressure to "load" the bearings.

After the differential bearing preload is set, the ring gear to pinion backlash should be adjusted. If the old ring gear and pinion is to be used, the backlash should be adjusted to the old setting to prevent any gear noise during operation. When a new gear set is used, mount a dial indicator in a position that can read against the drive or coast side of the ring gear teeth. Secure the pinion and rock the ring gear back and forth against the pinion teeth and note the indicator reading. Typically the reading should be 0.010 inch to 0.020 inch, but the actual specification should be followed. If the reading is above or below the specification loosen one adjuster ring one notch and tighten the other by one notch. Repeat this process until the proper specification is obtained. By rotating the adjuster rings equally, the bearing preload should not be affected. Check the gear tooth contact pattern for the contact location. By adjusting the backlash slightly, the pattern can be moved along the gear tooth.

Task D13   Check ring and pinion gear tooth contact pattern; interpret pattern, and adjust to manufacturer's specifications.

With the axle differential assembled, use a marking compound to paint at least twelve teeth of the ring gear. After rolling the ring gear, examine the marks left by the pinion gear contacting the ring gear teeth. A correct pattern is one that comes close to, but does not touch, the ends of the gear. As much of the pinion gear as possible should contact the ring gear tooth face without contacting or going over any edges of the ring gear. Adjusting the pinion gear will affect the contact pattern between the toe and heel, while adjusting the ring gear will affect the pattern from the top land to the root.

Task D14   Inspect, adjust and replace ring gear thrust block/bolt.

A thrust screw is incorporated into some axle differentials to allow the axle to withstand more torque. The thrust screw is designed to be a "stop" if the torque demanded from the ring and pinion ever causes the ring gear to deflect. If any deflection occurs, the thrust screw prevents the ring gear from deflecting to a point of tooth slip, which would cause extensive damage to the drive axle. To adjust the thrust screw, simply thread the screw into the axle housing until it rests on the ring gear. Then, back the thrust screw off one half turn and tighten the jam nut. The thrust screw is not designed to be in constant contact with the ring gear, serving only as a back-up measure for extreme torque conditions.

Task D15   Inspect, adjust, repair, or replace planetary gear-type two-speed axle assembly including case, idler pinion, pins, thrust washers, sliding clutch gear, shift fork, pivot, seals, cover, and springs.

A two-speed axle is similar to the double reduction axle except that a two-speed axle can be operated in either the single reduction or double reduction mode. Both are similar in design in that they use a planetary gear set as the second reduction gear set. An air or electric shift unit is mounted on the carrier and a switch in the cab of the vehicle actuates it. The driver can shift the drive unit into high or low gear range by actuating the switch. The air shift unit stops the driver from downshifting to a lower gear before the transmission is at a safe speed, preventing any possible damage from improper downshifting. Damage can still occur to the shift mechanism during upshifting if the driver is not paying attention to the conditions in which the vehicle is being operated.

Task D16   Inspect, repair, or replace air and electric two-speed axle shift control switches, speedometer adapters, motors, axle shift units, wires, connectors, hoses, and diaphragms.

Although some vehicles are equipped with electrical shift units, most axles are equipped with pneumatic shift systems. There are two air-activated shift system designs predominantly used to select the range of a dual range tandem axle or to engage a differential lockout. Ensure that air is present at the shift unit. These units tend to leak air into the axle housing when the seals or diaphragms leak. When internal air leaks are present or there is no movement of the shift forks with air applied, the complete shift unit should be replaced. Electric shift components can usually be checked with a volt/ohmmeter. Check for voltage into and out of the switch when the switch is in the on position. Check for voltage at the shift motor. If power is present at motor and the motor is not operating, disconnect the power from the motor and check the motors resistance. If the motor is open or shorted, the motor of the shift unit can sometimes be replaced separately but usually complete unit replacement is necessary. All electrical connectors should be cleaned with contact spray and sealed well when reconnected.

Task D17   Inspect, or replace power divider (interaxle differential) assembly.

The interaxle differential is an integral part of the front rear axle differential carrier in a tandem drive truck. Components of the interaxle differential are the same as in a regular differential: a spider (or cross), differential pinion gears, a case, washers, and the side gears. In an interaxle differential, the side gears perform a different job than the side gears of a standard differential. They transfer power to both the front and rear drive axles. Additional gearing at the front and rear drive axles splits this power and delivers it internally to the forwardmost drive axle and uses an output shaft and yoke to transmit power to the propeller shaft, which delivers power to the rearward drive axle.

Task D18   Inspect, adjust, repair, or replace air-operated power divider (interaxle differential) lockout assembly including diaphragms, seals, springs, yokes, pins, lines, hoses, fittings, and controls.

Most power dividers have a lockout mechanism that prevents the interaxle differential from splitting torque between the front and rear axles. The lockout mechanism consists of an air-operated lockout unit, a shift fork and pushrod assembly, and a sliding lockout clutch. When activated, the lockout unit extends the pushrod and shift fork. When the driver returns the activation switch to normal, allowing differential action between the axles, the return spring inside of the air shift unit draws the shift fork back to the unlocked position.

Task D19   Inspect and measure drive axle housing mating surfaces and alignment; determine needed repairs.

The differential carrier assembly can be steam cleaned while mounted in the housing as long as all openings are tightly plugged. Once removed from its housing, do not steam clean the differential carrier or any axle components. Steam cleaning at this time could allow water to be trapped in cored passages, leading to rust, lubricant contamination, and premature component wear. Once the axle housing and differential are properly cleaned, inspect for any signs of cracking and check the mating surfaces for notches, visible steps, or grooves. Most damage done to the differential and axle housing mating surfaces is caused by poor assembly practices. Most of this type of damage can be repaired by filing or slightly grinding the surface smooth and using an appropriate sealant.

Task D20 Inspect, service, or replace drive axle lubrication system components, pump, troughs, collectors, slingers, tubes, and filters.

Drive axle lubrication system components are, in most cases, a splash feed type system. In such systems the lubrication level is at a point where vital components are lubricated by the simple action of rotating parts coming into contact with the lubricant and then throwing off that lubricant as they are spun around. Additionally, some axles incorporate a pump and hoses or tubes to disperse the lubricant to critical parts in the axle differential. Always remember that passages need to be kept clear. Usually, checking the pump for smooth operation and forcing air through the passages during rebuilding is sufficient.

Task D21 Inspect and replace drive axle shafts.

For longer life, the surfaces of axle shafts are case hardened for wear resistance. A lower-hardness, ductile core is retained for toughness. Fatigue failures can occur in either or both of these areas. Failures can be classified into three types that are noticeable during a close visual inspection: surface, torsional (or twisting), and bending. Overloading the truck beyond the rated capacity or abusive operation of the truck over rough terrain generally causes fatigue failures.

Task D22 Remove and replace wheel assembly; check drive axle wheel/hub seal and axle flange gasket for leaks; determine needed repairs.

There are slight differences in bearing and seal service between grease- and oil-lubricated systems and front and rear drive axles. On a spoke wheel, the brake drum is mounted on the inboard side of the wheel and held in place with nuts. Servicing inboard brake drums on spoke wheels involves removing the single or dual wheel and drum as a single assembly. This involves removing the hub nut and disturbing hub components, so bearing and seal services are required. On disc wheels, the brake drum usually is mounted on the outboard side of the disc hub. The drum fits over the wheel studs and is secured between the wheel and hub. This means the wheel and drum can be dismounted without disturbing the hub nut. Outboard drums can be serviced without servicing the bearings and seals.

The wheel bearings are important components of the rear axle housing. The seals that protect the bearings can fail and allow oil to leak, thus causing damage to other components. Under certain circumstances, the road dirt or road water may enter the seal and cause more extensive damage. If possible seal damage is suspected, the bearing(s) should be removed, cleaned, inspected, and replaced if damaged. Always use a new seal whenever the housing or the bearings are removed. Be sure to check the housing vent and refill with the proper type and amount of lubricant after service.

Task D23 Diagnose drive axle wheel bearing noises and damage; determine needed repairs.

Wheel bearing noise is usually more prevalent under loaded conditions. With some vehicles, turning to the right and left can amplify bearing noise from the outboard hub during a turn. Noise will vary from faint clicks to deep rumbling, as the bearing wear gets more severe. Bearing service should be performed at the slightest sign of noise or looseness.

Problems associated with different models of axles and types of gearing can be specific to one model only. However, in most cases, one problem area generally can be caused by the same malfunction. The technician must bear in mind that universal joints, transmissions, tires, and drivelines can create noises that are often blamed on the drive axles. Experience and a keen ear are two of the most valuable tools in diagnosing driveline noises. Keep a record of past repair information to help diagnose future problems that are not obvious.

Task D24   Clean, inspect, lubricate, and replace wheel bearings; replace seals and wear rings; adjust drive axle wheel bearings.

Under normal operating conditions, axle wheel bearings are protected by lubricant carried into the wheel ends by the motion of axle shafts and gearing. Lubricant becomes trapped in the cavities of the wheel end and remains there, ensuring that lubricant is instantly available when the vehicle is in motion. In cases where wheel equipment is being installed, either new or after maintenance activity, these cavities are empty. Bearings must be manually supplied with adequate lubrication or they will be severely damaged before the normal motion of gearing and axle shafts can force lubricant to the hub ends of the housing. Improper wheel bearing endplay can result in looseness in the bearings or steering problems. When properly adjusted, the hub and wheel should rotate freely without excessive endplay.

The TMC (Truck Maintenance Council) has a recommended procedure (RP 618) for wheel bearing adjustment that is industry accepted. The procedure is as follows: Lubricate the bearings with clean oil of the same type as used in the drive axle or wheel hub. Torque the adjusting nut to 200 ft-lbs and rotate the wheel and recheck torque a couple of times to seat all components. Back off the adjusting nut until it is finger loose. Torque the adjuster nut to 50 ft-lbs while rotating the wheel. Back off the adjuster nut a specified amount as listed on TMC charts (⅙ to ½ turn depending on the nut(s) used). While again referencing the TMC chart Torque the jam or lock nut to a specified value. Check for bearing endplay of .001 inch to .005 inch with a dial indicator. If the endplay is not within this operating range the procedure must be performed again. This procedure is for adjustable wheel bearing hubs and should not be followed when servicing preset or unitized hubs. To ensure that drive axle hubs receive the correct amount of lubrication before operation, jack the opposite side of the axle to allow axle lube to flow out to the serviced hub. The axle should then be lowered onto a level surface and the axle fluid level should be checked and adjusted to the proper level.

Task D25   Inspect and test drive axle temperature sending unit/sensor; determine needed repairs.

The most reliable way to test a drive axle temperature sensor (thermistor type) for accuracy is to obtain a temperature-to-resistance chart from the manufacturer. This allows the technician to determine if the temperature sensor is sending a signal appropriate for the temperature it is encountering. Another way of diagnosing the temperature indicating circuit is to place a variable resistance in the place of the sensor. Varying a known resistance and checking the temperature display for correspondence is also a good way of helping pinpoint the source of the problem.

Task D26   Check, adjust, and replace wheel speed sensor(s).

Wheel speed sensors are very simple looking but rather sophisticated components. The wheel speed sensor produces an alternating current voltage signal that is sent to the control module. The wiring needs to be of the absolute best integrity to be a reliable conductor for the AC signal to accurately reach the module. A break in the wire at the wheel can actually cause total module failure by drawing water up to the connector pins on the module. When a break in a speed sensor wire occurs the sensor must be replaced and not repaired due to shielding concerns.

Wheel speed sensor adjustment is usually a simple procedure of pushing the sensor inward until it contacts the reluctor or tooth wheel. The rotation of the wheel will self adjust the sensor. Rust buildup or dirt on the reluctor wheel and improper wheel hub and drum installation are common causes of sensors being out of adjustment. The sensor can be checked for its resistance and output voltage. To test the sensor output voltage, block the wheels to prevent movement of the truck, raise the wheel of the sensor to be tested and release the park brakes. Disconnect the sensor and connect a DMM (digital multimeter) on the AC volts scale to the sensor terminals while the wheel is being rotated at approximately 30 rpm. The voltage should be at least 0.2 volts AC. While the sensor is disconnected set the DMM to the ohms/resistance scale. The resistance reading should be 700–3000 ohms. If the resistance reading is not within this range, the sensor must be replaced. If the voltage is lower than specified and the adjustment is correct, the sensor must be replaced.

Task D27   Clean, inspect, lubricate, and replace wheel end locking hubs.

Wheel end locking hubs require periodic inspection due to the chance of moisture and dirt working into the housing past its seal. The seal should be replaced and the hub should be repacked with grease when this condition is discovered. The hub must be shifted into low range and relies on a shift collar return spring to return it to high range. The hub may require replacement if the spring breaks and the shift collar is damaged from improper engagement.

Task D28   Inspect or replace extended service (sealed, close-tolerance, and unitized) bearing assemblies; perform initial installation adjustment procedures.

Wheel bearings are responsible for supporting the vehicle weight while allowing the wheel assembly to rotate. The proper adjustment requires some endplay or clearance for lubrication and bearing expansion at operating temperatures. The adjustment and proper lubrication is crucial for these bearings to operate without damage or seizure from excessive heat build-up. Some manufacturers are now using hubs that use bearing assemblies that produce preset clearances or sealed hub assemblies that reduce the chance of incorrect adjustment and lubricant. These hubs are retained by two bearing nuts and tabbed washer located between them. The inner nut is torqued to 500–700 lb.ft. for proper hub retention. The outer nut is torqued to 200–300 lb.ft to jam the inner nut and prevent any backing off. The tabs on the tabbed washer are bent in over the inner nut to hold it in place. When correctly installed these hubs should have 0.003 inch or less bearing play. If the bearing play is 0.006 inch or greater, rotational roughness, vibration, noise or seal leakage is present the hub must be replaced.

82. Technician A says that shift mechanism of a planetary double reduction final drive can be diagnosed using the same steps as the shift mechanism of a locking differential. Technician B says that on some models the second gear reduction comes from two helically cut gears. Who is right?
    A. A only
    B. B only
    C. Both A and B
    D. Neither A nor B   (D15)

83. Which of the following is the LEAST-Likely to be checked when replacing driveshaft U-joints?
    A. Driveshaft yoke phasing
    B. Final drive operating angle
    C. Driveshaft tube damage
    D. Final drive yoke damage   (C2)

84. What is the result of a completely failed interaxle differential lockout?
    A. The engine is not able to propel both the front and rear drive axles.
    B. The engine is not able to propel only the front axle.
    C. The engine is not able to propel only the rear axle.
    D. Difficult shifting from high to low speed.   (D18)

85. When replacing a flywheel ring gear, which of the following is correct procedure?
    A. Cool the ring gear in a freezer overnight.
    B. Heat the ring gear in an oven to 400° F (204° C).
    C. Cool the ring gear and heat the flywheel.
    D. Heat the ring gear and cool the flywheel.   (A11)

86. A driver complains of a clunking in the driveline at low speeds. Technician A says that this is likely a worn universal joint. Technician B says that it is likely a dry, underlubricated universal joint. Who is right?
    A. A only
    B. B only
    C. Both A and B
    D. Neither A nor B   (C1)

87. Technician A says that transmission mounts are used to absorb torque from the engine. Technician B says that transmission mounts will absorb drivetrain vibration. Who is right?
    A. A only
    B. B only
    C. Both A and B
    D. Neither A nor B   (B9)

88. When inspecting an input shaft for wear, which of the following is LEAST-Likely to be inspected?
    A. The front bearing retainer
    B. The output bearing
    C. The pilot bearing
    D. The input bearing   (B13)

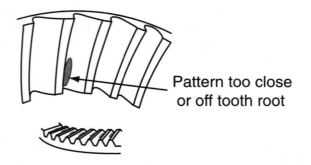

Pattern too close
or off tooth root

89. The tooth contact pattern shown in the figure is incorrect. What would have to be done to correct it?
   A. Move the pinion outward away from the ring gear.
   B. Move the ring gear closer into mesh with the pinion gear (decrease backlash).
   C. Move the pinion inward toward the ring gear.
   D. Move the ring gear away from the pinion gear (increase backlash).   (D11)

90. When installing a new differential carrier into an axle housing, a technician should check for all of the following **EXCEPT:**
   A. axle housing mounting flange for nicks scratches and burrs.
   B. damaged axle housing bolt holes or studs.
   C. carrier mounting flange for nicks scratches and burrs.
   D. run out of the ring gear.   (B18)

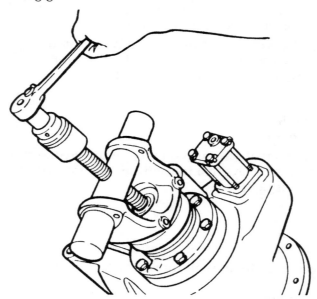

91. In the figure shown, the technician is:
   A. straightening the yoke.
   B. pressing the output shaft seal into place.
   C. removing the yoke.
   D. installing the yoke.   (C2)

92. How would a technician replace a ring gear once the rivets are removed?
   A. Press out the old one, heat the new ring gear in water, and assemble.
   B. Simply allow the old ring gear to separate from the differential case and install the new one.
   C. Pry the old ring gear off the differential and install the replacement ring gear with a press.
   D. Lightly hammer the old ring gear off the differential case and use a torch to heat the differential case before installing the new ring gear.   (D9)

93. A proper gear tooth contact pattern will:
    A. slightly overrun the toe of the gear.
    B. slightly overrun the heel of the gear.
    C. come close to, but not touch any edge of the gear tooth.
    D. come to within ½ of an inch of the gear toe.   (D13)

94. What would produce a broken synchronizer in the auxiliary of a twin countershaft transmission?
    A. A faulty interlock mechanism.
    B. Improper driveline angularity.
    C. A twisted mainshaft.
    D. Improper towing of the truck.   (B18)

95. What should a technician do when servicing the gears of a differential equipped with a lubrication pump?
    A. Replace all internal hoses or lines.
    B. Pack the pump full of lithium-based grease to ensure priming of the system after installation.
    C. Replace all external hoses.
    D. Check the pump for smooth operation and blow forced air through all passages.   (D20)

96. What would cause a hard or stiff shift in or out of third gear in a twin countershaft transmission?
    A. A faulty air pressure regulator
    B. A twisted mainshaft
    C. A worn synchronizer
    D. A faulty clutch brake   (B14)

97. When measuring pinion bearing preload:
    A. note the shim size used and use the same size during installation.
    B. note the shim size used and choose one that is 0.001 inch (0.025 mm) smaller for installation.
    C. note the shim size used and choose one that is 0.001 inch (0.025 mm) larger for installation.
    D. never use the same shim to test preload and install in the axle.   (D10)

98. A technician is measuring differential bearing preload. After following the proper procedure, the dial indicator reads "0." What should the technician do next?
    A. Loosen the adjusting ring one notch.
    B. Continue with assembly; the preload is set correctly.
    C. Tighten each bearing adjusting ring one notch.
    D. Adjust the bearing preload of the other side of the differential.   (D10)

99. A flywheel has a damaged pilot bearing. Technician A says that damage could be caused by poor maintenance habits. Technician B says that damage could be caused by bell housing misalignment. Who is right?
    A. A only
    B. B only
    C. Both A and B
    D. Neither A nor B   (A9)

100. A truck driver complains that he cannot shift out of interaxle differential lock. Which of the following could be the cause?
    A. A broken shift shaft spring
    B. A broken shift shaft
    C. An open or damaged air line
    D. A stripped differential clutch collar   (D18)

101. Technician A says that a broken wheel speed sensor wire can be repaired with a crimp splice or equivalent. Technician B says that you must replace the entire wheel speed sensor to repair the condition. Who is right?
    A. A only
    B. B only
    C. Both A and B
    D. Neither A nor B   (D26)

102. In a twin countershaft transmission, a noise is noticeable in all gear shift positions except for high gear (direct). The Most-Likely cause of this noise is:
    A. a worn countershaft gear.
    B. worn countershaft bearings.
    C. worn rear main shaft support bearings.
    D. worn front main shaft support bearings.   (B1)

103. A truck with an electronically automated mechanical transmission is in the shop for repairs. Technician A says that with an electronic service tool (scan tool) you can retrieve transmission information and operating data. Technician B says that the electronic service tool can change gear ratios. Who is right?
    A. A only
    B. B only
    C. Both A and B
    D. Neither A nor B   (B6)

104. Under what conditions would a failed pilot bearing be the most noticeable?
    A. Under heavy acceleration in first or reverse.
    B. During steady cruising at highway speeds.
    C. Standing still with the engine off while moving the clutch pedal.
    D. Sitting at a red light in gear.   (A9)

Appendices

Answers to the Test Questions
for the Sample Test Section 5

1.	A	16.	A	31.	A	46.	C
2.	D	17.	A	32.	D	47.	B
3.	B	18.	C	33.	C	48.	C
4.	C	19.	C	34.	C	49.	B
5.	C	20.	A	35.	C	50.	A
6.	B	21.	C	36.	D	51.	B
7.	A	22.	D	37.	D	52.	A
8.	C	23.	B	38.	D	53.	C
9.	C	24.	A	39.	D	54.	D
10.	A	25.	D	40.	C	55.	B
11.	B	26.	B	41.	A	56.	B
12.	C	27.	C	42.	D	57.	B
13.	C	28.	B	43.	D	58.	A
14.	B	29.	A	44.	B	59.	C
15.	A	30.	B	45.	A	60.	C

Explanations to the Answers for the Sample Test Section 5

Question #1

Answer A is correct because a damaged gasket or missing sealant could cause a leak in this location. Answer B is wrong because overloading of the drive train would increase the axle temperature and thin out the fluid but a good gasket should still seal this area.

Answer C is wrong because a plugged breather would increase the pressure inside the axle assembly, but this condition usually force oil past the lips of the wheel seal which is the weakest sealing point.

Answer D is wrong because moisture contamination will have no effect on the ability of gaskets to seal or leak.

Question #2

Answers A, B, and C are wrong because these are all valid mechanical reasons for causing the front axle to be nonpowered. A faulty de-clutch will prevent the power flow from reaching the front axle assembly. A broken differential will prevent drive to the axle shafts and a stripped ring gear will prevent the pinion from driving the differential case.

Answer D is correct because the unit is not electrically operated and therefore the LEAST-Likely cause.

Question #3

Answer A is wrong because the measurement is within specification and no resurfacing is needed.

Answer B is correct because the specification is 0.006 to 0.015 and no service is required.

Answers C and D are wrong because flywheel servicing and replacing the pilot bearing will have no effect on the flywheel-housing bore. These are both flywheel service conditions.

Question #4

Answer C is correct because Technicians A and B are both right. A plugged oil line can cause poor differential lubrication and therefore heat buildup from friction. Failure to lock the inter axle differential during slippery road conditions can allow a wheel spinout condition on one axle which will cause extreme differential gear speeds and the generation of intense heat.

Question #5

Answers A and B are wrong because a faulty master cylinder piston seal or a minute line leak will allow apply pressure to bleed off which will allow the clutch to slowly re-engage.

Answer C is correct because it is the LEAST-Likely cause. Binding linkage would allow the truck to creep forward at all times because of incomplete disengagement.

Answer D is also wrong because a leaking or weak air servo cylinder may also allow a slow re-engagement due to hydraulic pressure loss or air pressure bypassing the servo piston.

Question #6

Answer B is correct. Technician B is correct because adjusting the release bearing to clutch brake clearance will affect the position of the release bearing. The clutch pedal free play should only be adjusted after the release bearing is in its correct position, which makes Technician A incorrect.

Question #7

Answer A is correct. Technician A is correct because it is a recommended practice to flush the transmission cooler system when the transmission is being rebuilt. This prevents any foreign particles from being circulated to and from the cooler possibly damaging the freshly rebuilt transmission.

Question #8

Answer A is wrong because broken teeth on the forward drive axle ring gear can cause the front axle to be nonpowered.

Answer B is wrong because broken teeth on the rear drive axle ring gear would not affect the power flow to the rear axle.

Answer C is correct because stripped output shaft splines would still allow power to reach the front drive axle but the interaxle differential side gear would slip on the output shaft, producing no drive to the rear drive axle.

Answer D is wrong because interaxle differential damage could render both axles powerless.

Question #9

Answers A and D are wrong because there are usually no dipsticks or sight glasses on a manual transmission. Answer B is wrong because even though you can feel lubrication it does not mean it is at the proper level.

Answer C is correct because when the oil level is even with the bottom of the filler plug hole it is at the recommended level. The bottom of the filler plug is positioned at the manufacturer's recommended fluid level.

Question #10

Answer A is correct because a short to ground on the negative wire to the lamp will ground the circuit and cause the lamp to stay on constantly.

Answer B is wrong because a positive short to the negative wire will allow no voltage differential across the bulb, causing no illumination.

Answer C is wrong because if the switch will not close, the circuit will not be able to be completed.

Answer D is wrong because a short to ground on the positive wire will not allow voltage to reach the light bulb. Both answers B and D would also cause blown fuses.

Question #11

Answer B is correct. When discussing wheel bearing service, Technician A is wrong because wheel bearings should only be replaced when wear or damage warrants replacement. Technician B is right because raising the opposite side of the axle is the correct procedure for filling the bearing cavity with axle lube. Pre-filling the hub before installation will not place enough fluid in the hub.

Question #12

Answer C is correct. Technicians A and B are correct. A brinelled trunnion will tend to make clicking sounds during operation and a bearing that is separated from its bearing plate can make noises due to the change of movement during operation.

Question #13

Answer C is correct. Technicians A and B are correct. The cone clutching surface is crucial for synchronizer operation. This surface must have ridges to cut through the transmission fluid to enable it to grip the gear. Wear of these surfaces is usually caused by incorrect driving habits, such as not using the clutch during gear shifting.

Question #14

Answer B is correct. Technician A is wrong because this valve cannot be disassembled and cleaned and reinstalled. Technician B is right because this air shift valve is not serviceable and must be replaced as a unit. The cost and time to disassemble many air valves, obtain parts and the light aluminum bodies make rebuilding impractical.

Question #15

Answer A is correct. Technician A is right because to prevent driveline vibrations and speed oscillations that could produce U-joint failure, the joints should operate within one degree of each other. Technician B is wrong because three degrees of difference will produce an unequal speed fluctuation between the transmission and the final drive, which can produce a binding sensation and vibration in the driveline.

Question #16

Answer A is correct. Technician A is correct. The wear compensator is spring loaded and must be removed from mesh with the adjuster ring prior to manually rotating it. Technician B is wrong because this adjustment would not be possible or compensator damage would occur.

Question #17

Answer A is correct because the spring seats are the recommended jack points on the rear axle. The spring seats are generally flat and place even loading on the axle housing preventing housing damage. Answers B, C, and D are wrong because these placements can cause damage to the axle housing or slipping conditions that could allow the axle to fall and personal injury to occur.

Question #18

Answer A is wrong because worn bearings can allow shaft movement, which may allow the transmission to jump out of gear.
Answer B is wrong because faulty detents can allow the shift rails to move freely causing jumping out of gear.
Answer C is correct because broken gear teeth may make noise, but typically will not cause the transmission to jump out of gear.
Answer D is wrong because faulty engine mounts can cause the transmission to jump out of gear due to transmission movement.

Question #19

Answer C is correct. Technicians A and B are correct. A special data link tester is the only tool that is capable of checking for data link communication. It is capable of receiving and sending test data for operation verification. Both the data link tester on its continuity setting and an ohmmeter are capable of testing for data link continuity.

Question #20

Answer A is correct because you would only remove the instrument panel gauge and test for proper movement after the sensor and wiring were ruled out as possibilities.
Answers B, C, and D are wrong because installing a testing resistor, checking sensor resistance, and checking connector conditions are all good first choices for a technician who needs to diagnose the problem.

Question #21

Answer A is wrong because some flywheels require replacement if ring gear teeth are damaged.
Answer B is wrong because you can remove the flywheel and install a new ring gear as an acceptable repair.
Answer C is correct because Mig welding new teeth on is the LEAST-Likely, and not a proper repair.
Answer D is wrong because sending the flywheel out to a jobber for repair can be an appropriate repair.

Question #22

Answer D is correct because the pressure plate is being checked for warpage. Warpage will cause chatter and this is most noticeable at takeoff due to high torque demands.
Answers A and B are wrong because when the pressure plate is fully engaged, chatter will not usually be noticeable.
Answer C is wrong because torque demand is less at high speeds and chatter will not be as noticeable.

Question #23

Answer B is correct. Technician A is wrong because any pinion related noises would be noticeable under most driving conditions and not only when cornering. Technician B is correct because a noise that is noticeable when cornering is usually related to differential gearing. These gears only move relative to each other when one axle must rotate at a different speed than the other.

Question #24

Answer A is correct because 0.005 to 0.012 inch (0.127 to 0.305 mm) is the correct measurement. Answer B and C are wrong because these measurements are excessive and the shaft movement that would be allowed could allow gear, bearing, and retainer damage.

Answer D is also wrong because the measurements are insufficient and thrust bearing damage could occur as the shaft expands during operation.

Question #25

Answer A is wrong because excessive clutch brake usage will only try to slow down the input shaft. This would Most-Likely cause premature clutch brake failure.

Answer B is also wrong because during assembly is the Most-Likely time that the countershafts would be incorrectly timed. The transmission would not be able to turn at all in this situation.

Answer C is also wrong because if a truck is towed with the axles left in place, the driveshaft will turn the main shaft causing bearing damage due to lack of lubrication.

Answer D is correct because shocks from the driveline can cause this type of damage due to excessive twisting forces.

Question #26

Answer A is wrong because too little clutch pedal free play would affect disengagement, not engagement.

Answer B is correct because binding linkage can prevent the pressure plate from applying its full clamping force causing slippage. Clutch disc slippage can cause burning of the clutch lining surface.

Question #27

Answer C is correct. Technicians A and B are correct. Whenever possible a universal joint press or a puller should be used to remove a universal joint from a yoke but, it is also acceptable to use a hammer on a yoke that is supported by the cross in a vice.

Question #28

Answer B is correct. Technician B is correct. A unitized wheel hub is an assembly that is torqued into place. This assembly contains both inner and outer wheel bearings and races as well as the wheel seal. It has provisions for free play built into it.

Question #29

Answer A is correct because the ⅛ inch of clearance at the release bearing produces 1½ inch to 2 inches of clutch pedal free play.

Answer B is wrong because ½ inch of clearance would produce excessive pedal free play.

Answers C and D are wrong because push type clutches do not use clutch brakes.

Question #30

Answers A, C, and D are wrong because a loose end yoke, worn slip yoke spline, and a bent driveshaft tube can cause PTO driveshaft vibration. All of these factors can allow radial movement or wobble in the driveshaft, which will create vibrations.

Answer B is correct because an out-of-balance driveshaft is the LEAST-Likely cause of a PTO vibration.

Question #31

Answer A is correct because a dirty or plugged air filter will cause the transmission to not shift into high range due to insufficient airflow.

Answer B is wrong because the figure shown is not an electric shift unit.

Answers C and D are wrong because a worn range synchronizer and wear on the gear teeth will affect shifting quality, not render the system unable to shift.

Question #32
Answer D is correct because both technicians are wrong. Technician A is wrong because it is not good practice to replace parts before finding the fault. Technician B is wrong because most sensors need at least 5 volts to operate properly.

Question #33
Answer A is wrong because excessive bearing wear is unusual for a vehicle with such low mileage.
Answer B is wrong because transmission fluid does not evaporate naturally.
Answer C is correct because a plugged transmission breather filter causes excessive internal pressure due to expansion of the air and fluid in the transmission case as they heat up and expand during operation.
Answer D is wrong because a poor quality filter would not affect fluid leaking past seals.

Question #34
Answer C is correct. Technicians A and B are correct. Loose spring U-bolts can allow movement of the rear axle housing which will affect driveline angles. Axle shims that are left out or reversed will also affect the driveline angles which can create vibrations.

Question #35
Answer C is correct. Technician A is correct because for balance and phasing purposes, shafts should be reassembled in the same manner as they were before they were serviced. Technician B is also correct because if a weight is removed it could affect the shafts balance, which could cause a vibration.

Question #36
Answer D is correct because neither technician is right due to the fact that the figure shows a correct tooth pattern for new gear sets. As the teeth break in, the pattern should move away from the toe toward the center of the tooth.

Question #37
Answer D is correct.
Answers A, B, and C are wrong because the solenoids in the range valves when energized block off the exhaust ports and allow air flow to the appropriate range cylinder. If the vehicle is at rest the solenoids are de-energized and both sides of the range cylinder are open to exhaust. This will allow the range cylinder to remain in the position of the last selection.

Question #38
Answer A is wrong because oil on the friction disc provides a lubricant between the friction disc and the pressure plate which will allow slippage which generates heat and can burn the pressure plate.
Answers B and C prevent complete clutch engagement, which will decrease the pressure plate's apply force and allow disc slippage and pressure plate burning.
Answer D is correct because the pilot bearing will have no effect on the friction material or the clamping force of the pressure plate so no burning will take place.

Question #39
Answers A, B, and C are wrong because improper clutch use of this kind should not affect the collar clutches, input shaft, or the clutch linkage.
Answer D is correct because depressing the clutch pedal to the floor with each shift will force the release bearing against the clutch brake causing early clutch brake failure. The clutch brake is designed to stop or slow the input shaft for initial gear selection.

Question #40
Answer A is wrong because support bearings are sealed.
Answer B is wrong because there is usually no pressing necessary for support bearings.
Answer C is correct because you fill the entire cavity around the bearing with grease to protect the bearing from water and contaminants.
Answer D is wrong because orientation of the support bearing is not critical.

Question #41
Answer A is correct because the technician should replace the ring gear and pinion for a tooth spalling condition.
Answer B is wrong because the ring gear and pinion should always be replaced as a set.
Answer C is wrong because changing lubrication will not fix the problem of tooth spalling which will eventually damage the pinion gear.
Answer D is wrong because this does not fix the problem of tooth spalling; it will only correct the backlash.

Question #42
Answer D is correct because neither technician is right. The correct process is to turn the thrust screw until it stops against the ring gear, and then loosen the thrust screw one-half turn, and lock the jam nut. One full turn would allow too much ring gear side thrust and setting the block in contact with the ring gear would cause wear on the ring gear and scoring of the thrust block.

Question #43
Answer A is wrong because a defective inner pinion bearing tends to generate noise during acceleration because the pinion is being forced back against it by the pinion and ring gears' separation forces under load.
Answers B and C are wrong because differential gearing and bearings will only generate noise when they are rotating, which is during turns.
Answer D is correct because a defective outer pinion bearing will make noise on deceleration because the ring gear tends to pull the pinion inward placing the load against the damaged bearing.

Question #44
Answer A is wrong because a shift collar can only physically engage one gear at a time.
Answer B is correct because the interlock mechanism is designed to allow the movement of only one shift rail at a time.
Answer C is wrong because a broken detent spring will allow jumping out of gear only.
Answer D is wrong because an incorrectly installed shifter lever will Most-Likely prevent the selection of any gear.

Question #45
Answer A is correct because a broken road speed sensor is the LEAST-Likely cause when the odometer still functions.
Answers B, C, and D are wrong because loose wiring, a broken speedometer gauge, and an open circuit are all possible causes. The road speed sensor feeds both the odometer and speedometer, which indicates a problem with the speedometer circuit or gauge can be the only possibility for this condition.

Question #46
Answers A, B, and D are wrong because aligning the clutch disc, adjusting the release bearing, and lubricating the pilot bearing must be performed when installing a new pull-type clutch.
Answer C is correct because you would only resurface the limited torque clutch brake if it had surface damage or unevenness.

Question #47
Answer A is wrong because if Technician A has to bend a finger to feel the fluid, the level is not appropriate.
Answer B is correct because the fluid level must be even with the bottom of the fill hole.

Question #48
Answer C is correct. Technicians A and B are correct. The ½ inch measurement between the release bearing and clutch brake provides the necessary pedal travel to produce release bearing to clutch brake contact at the full pedal position. The ⅛ inch release fork to release bearing clearance produces the correct clutch pedal free play.

Question #49

Answer B is correct. Technician A is wrong because replacement of all seals is not necessary. Only any seals that show signs of damage should be changed. Technician B is correct because changing to a higher grade of transmission oil may be all that will be necessary. Only if a leak is present after the better oil is installed should any seals be replaced.

Question #50

Answer A is correct. Technician A is correct. Incorrect driveline angles can cause growls on acceleration and gearshift vibrations of deceleration. A computerized angle analyser program can help identify incorrect angularity. Technician B is wrong because vibrations and growls are generally driveline related.

Question #51

Answer B is correct. Technician B is correct. The clutch pedal must be able to be depressed fully to engage the clutch brake control valve. When the clutch brake control valve is operating, it allows air flow to the clutch brake. An obstructed clutch pedal would prevent this from happening.

Question #52

Answer A is correct. Technician A is correct. Hydraulic retarders operate by circulating oil into a cavity that contains a spinning rotor. The chance of rotor failure is rare because the rotor does not contact the housing. The air and hydraulic control circuits contain valves, lines, and hoses, which are susceptible to damage and failure.

Question #53

Answer C is correct because both technicians are right. A broken intermediate plate can be caused by poor driver technique or by pulling loads that are too heavy. Shock loads, overloading, and extreme heat generated by riding the clutch or slippage can cause pressure plate and intermediate plate cracks.

Question #54

Answer A is wrong because the pinion depth is set and should not be moved because the pattern is centered between the root and the top land of the tooth.
Answer B is wrong because the pattern is not correct.
Answer C is wrong because there is not enough backlash. Backlash determines the pattern's position between the toe and the heel of the tooth.
Answer D is correct because you adjust the bearing adjusting rings to increase the amount of backlash, which will move the pattern slightly up the tooth toward the heel.

Question #55

Answer A is wrong. Dirty transmission fluid would not leave any effect here because oil does not circulate between the outer bearing race and its bore.
Answer B is correct because these marks are signs of fretting. Normal transmission vibrations will cause fretting on the bearing cup, which is the transfer of the bores machining patterns to the bearing race.
Answer C is wrong because a spun bearing would cause a much worse mark to the bearing bore.
Answer D is wrong because this is not a sign of a poorly manufactured bearing.

Question #56

Answers A, C, and D are wrong because equalizing tire pressures, placing the transmission output yoke vertically, and placing the transmission in neutral should all be done before driveline angle measurement.
Answer B is correct because when a level surface is not available to park the truck on, using jack stands to level the vehicle is not acceptable. Leveling the truck by placing shims under the tires is acceptable.

Question #57

Answer B is correct. Technician A is wrong because this is not a reliable method of testing shift linkage. Interference in the shifter housing and movement of the shift collars could also cause resistance. Technician B is correct because you have to disconnect the linkage at the transmission to check the linkage independently from the transmission. The transmission shifting mechanism can also be checked when the linkage is disconnected.

Question #58
Answer A is correct because when the engine is idling and the clutch pedal is fully depressed, the clutch is released. At this point the transmission input shaft is turning inside the pilot bearing, not with it. The clutch is not keeping the input shaft centered and driving it at the same speed as the pilot bearing. Answers B, C, and D are wrong because all of these selections have the clutch engaged, which does not allow any speed difference between the pilot bearing and the input shaft.

Question #59
Answers A, B, and D are wrong because an inertia brake that uses an electric coil does not use air pressure, diaphragms, or accumulators for brake engagement.
Answer C is correct because the inertia brake does rely on friction and reaction discs to produce the countershaft braking.

Question #60
Answer A is wrong because a worn or rough release bearing would cause a rough pedal feel and noise upon engagement.
Answer B is wrong because excessive input shaft end play could cause vibration.
Answer C is correct because a leaking rear main seal will allow oil to contaminate the friction surface of the clutch and cause slippage.
Answer D is wrong because broken torsional springs would cause chatter upon clutch engagement.

Answers to the Test Questions
for the Additional Test Questions Section 6

1.	A	27.	B	53.	A	79.	A
2.	C	28.	A	54.	D	80.	D
3.	C	29.	C	55.	B	81.	A
4.	B	30.	C	56.	B	82.	C
5.	C	31.	B	57.	C	83.	B
6.	A	32.	A	58.	A	84.	A
7.	D	33.	D	59.	A	85.	B
8.	A	34.	B	60.	B	86.	C
9.	B	35.	A	61.	B	87.	C
10.	C	36.	A	62.	A	88.	B
11.	A	37.	A	63.	B	89.	A
12.	B	38.	C	64.	D	90.	D
13.	B	39.	A	65.	A	91.	C
14.	C	40.	A	66.	A	92.	A
15.	B	41.	A	67.	C	93.	D
16.	C	42.	B	68.	C	94.	B
17.	B	43.	A	69.	B	95.	D
18.	D	44.	C	70.	C	96.	B
19.	B	45.	B	71.	A	97.	C
20.	B	46.	D	72.	B	98.	C
21.	A	47.	A	73.	C	99.	B
22.	B	48.	D	74.	D	100.	A
23.	D	49.	B	75.	D	101.	B
24.	C	50.	B	76.	C	102.	D
25.	B	51.	B	77.	C	103.	A
26.	C	52.	B	78.	C	104.	D

Explanations to the Answers for the Additional Test Questions Section 6

Question #1
Answer A is correct because this is a normal condition. The oil is not being circulated and cooled. The hot components tend to heat up the oil for a short time period before it begins to cool down.
Answers B and C are wrong because restricted pump cooling circuit and plugged or blocked transmission cooler fins would be evident as overheating during operation.
Answer D is wrong because transmission fluid does not lose its thermal inertia.

Question #2
Answer C is correct. Technician A is correct because it is recommended procedure to repair a damaged crankshaft seal wear surface with a wear sleeve and appropriate seal. Technician B is also correct because some manufacturers allow for a deeper installation of the seal, which will place the seal lip at an area on the crankshaft that is not grooved. Because both technicians are right.

Question #3
Answers A and B are wrong because poor quality axle fluid and wheel bearing problems could cause overheating.
Answer C is correct because if the interaxle differential was not operating smoothly it should only affect the temperatures in the front drive axle.
Answer D is wrong because continuous vehicle overloading puts extra torque and force on the components, which generates more heat.

Question #4
Answer B is correct because the marks are used to time the main shaft to the countershafts on a twin shaft transmission.
Answers A, C, and D are wrong because there would be no purpose for marking the gears is this fashion to identify tooth wear, the number of teeth on the gears, or the type of gear.

Question #5
Answer C is correct. Both technicians are correct because twin disc clutches are used for high torque application and it also uses an intermediate plate between the two clutch discs. This intermediate plate doubles the surface area, which doubles the friction surface and the amount of torque the clutch can accept.

Question #6
Answer A is correct. Technician A is correct because drive pinion depth should be set only after you properly preload the pinion bearing. If the pinion is not properly adjusted the depth setting will be incorrect. Technician B is wrong because the drive pinion should be properly located before the ring gear is installed.

Question #7
Answer A is wrong because the clutch brake is used for slowing or stopping the input shaft while shifting into reverse or first gear.
Answer B is wrong because the clutch brake is used for reducing gear clash when shifting from gear to gear because of the slowing of the input shaft.
Answer C is wrong because the clutch brake is used for reducing gear clashing and damage during shifts.
Answer D is correct because the LEAST-Likely purpose of the clutch brake is reducing U-joint wear. The clutch brake has no effect on the U-joints other than reducing the shocks from poor clutch operation.

Question #8
Answer A is correct. Technician A is correct because as bearings wear and age their surface hardness may begin to flake or show signs of spalling due to the stressing, heat changes, and loads they operate with. Technician B is wrong because dirt tends to leave small dents and lines in bearings.

Question #9
Answers A and D are wrong because the procedures are incomplete. Air lines and yokes would need to be taken off.
Answer B is correct because you do have to disconnect the air line, remove the output shaft yoke, power divider cover, and all applicable gears as an assembly.
Answer C is wrong because it is not necessary to remove the entire differential carrier.

Question #10
Answers A and D are wrong because the gears are in constant mesh, so they are less susceptible to gear tooth wear, chipping, or breakage.
Answer B is wrong because idler gears are fixed to the idler shaft; therefore no rotation is possible to cause inner race spalling.
Answer C is correct because idler shaft bearing wear results from constant mesh and its speed of rotation.

Question #11
Answer A is correct because when the breather is plugged, internal pressure can build inside the transmission and push the oil out.
Answers B, C, and D are wrong because these are all exit routes for the pressurized oil, not the source of the pressure.

Question #12
Answers A and D are wrong because a few particles indicate normal wear, not a problem needing immediate resolution.
Answer B is correct because you do inform the customer of the condition and tell them to monitor the amount of particles.
Answer C is wrong because it is always best to inform the customer of any condition on their vehicle that may require extra attention.

Question #13
Answer A is wrong because flywheel-housing run out requires the dial indicator to be on the crankshaft.
Answer B is correct because in the figure shown the technician is checking for flywheel face run out.
Answer C is wrong because the dial indicator is not positioned for flywheel radial run out measurement.
Answer D is wrong because the dial indicator is not contacting the pilot bearing bore.

Question #14
Answers A, B, and D are wrong because these problems would all cause vibrations but they would only be noticeable whenever the PTO was operating, not when the vehicle was moving on the highway.
Answer C is correct because the driver's diagnosis must be incorrect, thinking that the vibration is being caused by the PTO. A vibration during low speed shifting would more likely be caused by driveline angle changes from component movement during high torque application.

Question #15
Answer B is correct. Technician A is wrong because the angle between the drive pinion centerline and a true vertical is the axle's installed angle. Technician B is correct because driveline angle measurement is angle formed between the transmission output shaft and the driveshaft centerline.

Question #16
Answer A is wrong because replacing the breather is not usually required.
Answer B is wrong because gasoline should not be used as a cleaning agent.
Answer C is correct because the best way to clean a transmission breather is to remove it and use solvent to break down and wash out sludge and dirt and then compressed air will help blow out the remaining dirt.
Answer D is wrong because wiping the orifice will not clean or remove any dirt trapped inside the breather.

Question #17
Answer B is correct because the technician is removing the front interaxle side gear bushing.
Answers A, C, and D are wrong because there the figure is showing a bushing being pressed from a gear and not installing a side bearing, setting a bearing race, or adjusting preload.

Question #18
Answer A is wrong because the actuator arm must be in the correct location for the clutch to be able to self-adjust.
Answer B is wrong because a bent adjuster arm can prevent self-adjustment.
Answer C is wrong because the adjusting ring must be free to rotate.
Answer D is correct because the pilot bearing is responsible for input shaft alignment and does not affect adjustment of a self-adjusting clutch.

Question #19
Answer B is correct. Technician A is wrong because there is no provision for adjustment in the pressure plates of push type clutches. Technician B is right because all of the adjustment is done through the linkage on a push-type clutch.

Question #20
Answer B is correct. Technician A is wrong because a special lithium-based grease will work for slip splines but is not required. Technician B is correct because good quality U-joint grease can also be used on slip splines.

Question #21
Answer A is correct because removing the transmission, then the torque converter, then the oil pump is the correct procedure.
Answers B and C are wrong because the transmission oil pump is not accessible through the oil pan; the transmission must be removed.
Answer D is wrong because the extra step of removing the bell housing is not necessary.

Question #22
Answer A is wrong because it is normal for drive axles to acquire slight condensation (which usually "boils off" during normal driving conditions) but not heavy condensation.
Answer B is correct because infrequent driving and short trips do not bring the axle to temperature and moisture does not boil off.
Answer C is wrong because although submerging the axles in water can introduce water into the axle, it is not the Most-Likely cause.
Answer D is wrong because any water that happens to make its way into the axle would normally be "boiled off" during normal driving conditions.

Question #23
Answers A, B, and C are wrong because usually driveshaft vibration will not be evident in the lower-to-moderate speed ranges whether under load or not.
Answer D is correct because when the truck is above 50 mph (80 km) with no load, the driveshaft is spinning nearest to its maximum speed range and centrifugal force will have its greatest effect on the imbalance. When the shaft is under load the extra stress will sometimes limit the effects of the imbalance.

Question #24
Answer C is correct.
Answers A, B, and D are wrong because bearing roughness, excessive bearing endplay, and a leaking wheel seal are all valid reasons for hub replacement. Unitized hubs are factory assembled sealed units that are only serviceable by replacement. Bearing play in a unitized wheel hub should not exceed 0.006 inch and because 0.003 inch is within that specification, this is the exception.

Question #25
Answer A is wrong because poor quality bearings would be evidenced by damage to the bearings themselves.
Answer B is correct; poor quality lubricant is the Most-Likely cause. The input gears will show damage before others because the input gears turn at all times.
Answer C is wrong because inferior quality parts would be evidenced by damage other than overheating.
Answer D is wrong because overloading of the drive train would be evident in many areas of the drive train, not just the input shaft of the transfer case.

Question #26
Answer C is correct.
Technician A is correct. Even though the wiper ring is a friction fit on the spindle shoulder, to prevent oil seepage, sealant should always be used. Technician B is also correct because the seal that is installed when a wiper ring is used has a larger inside diameter.

Question #27
Answer A is wrong because moisture is a product of compressing air. Small amounts of moisture can be tolerated in some air systems.
Answer B is correct because air-operated equipment depends on air pressure in only one direction; movement in the opposite direction requires mechanical means.
Answer C is wrong because most air systems need some method of filtering incoming air.
Answer D is wrong because air movement is not sufficiently affected by bends in the line.

Question #28
Answer A is correct. Technician A is correct because the pedal free travel should be about 1.5 to 2 inches (38.1 to 50.8 mm). This clearance ensures full clutch engagement and no interference from the linkage should occur. Technician B is wrong because if the release bearing to clutch brake travel is less than ½ inch the release bearing will contact the clutch brake to soon causing clutch brake damage and complete clutch disengagement may not be possible.

Question #29
Answers A, B, and D are wrong because small cracks, bends, and twists are all valid reasons for replacement of the axle shaft. All of these conditions are signs of axle shaft damage.
Answer C is correct because pitting of the axle shaft is not a valid reason to replace the axle shaft, therefore the exception. Pitting can be present on new shafts as well as old shafts and is not considered harmful to shaft operation.

Question #30
Answer C is correct. Both technicians are correct because pre-set hubs are adjusted by simply torquing to specification and these hubs are designed to operate with minimal free play.

Question #31
Answers A, C, and D are wrong because an extended high torque situation, high temperature operation, and poor lubricant quality may be causes of bearing failure, but not leading causes.
Answer B is correct because dirt in the lubricant is the most common cause of transmission bearing failure. Dirt is present in all transmissions and it can be very abrasive. It is estimated that more than 90 percent of all bearing failures are dirt related.

Question #32

Answer A is correct because the pilot bearing stub seldom cracks and this is the LEAST-Likely damage to be sustained by an input shaft.

Answers B, C, and D are wrong. Checking for gear teeth damage, input spline damage, cracking, or other fatigue wear are all valid checks because these are regularly occurring damage conditions.

Question #33

Answers A, B, and C are wrong because digital multimeters, laptops, and hand-held scan tools are commonly used by technicians for diagnosing electronically automated transmissions.

Answer D is correct because a test light would be the LEAST-Likely tool used for diagnosing. Test lights should not be used when diagnosing because these are not high impedance tools and may cause damage to sensitive electronic components.

Question #34

Answer A is wrong because it is not necessary to fill the gouges. Additional sealant is sufficient to seal most gouges that could be present.

Answer B is correct because a proper job requires you to grind and sand smooth any imperfections. Some nicks may still require additional sealant.

Question #35

Answer A is correct because binding bushings will affect linkage operation. Binding bushings can cause stiff and jerky operation, which could result in hard or no shifting.

Answer B is wrong because linkage length is only a factor at initial installation.

Answer C is wrong because linkage can have bends or twists; this can be a normal part of the linkage.

Answer D is wrong because for the most part, surface rust and pitting will not affect the functionality of linkage.

Question #36

Answer A is correct. Technician A is correct because small dirt particles can cause pitting and bearing surface wear and small metal particles can wedge in the races and cause a bearing to spin in its bore. Technician B is wrong because a low fluid level would tend to affect the upper bearings before a lower bearing would sustain such damage.

Question #37

Answer A is correct because on disassembly you look for shims during removal of the old bearing to maintain the original position.

Answer B is wrong because the sealed bearings come lubricated from the manufacturer. The outer cavity around the bearing should be filled with grease to prevent the possible entry of dust and moisture into the bearing.

Answer C is wrong because driveline angle adjustment should not be necessary if the procedure is followed and the correct bearing is installed.

Answer D is wrong because installation with hand or air tools should not have any affect on quality.

Question #38

Answer C is correct. Both technicians are right. A fault code can be retrieved from the service light or by a hand-held scan tool.

Question #39

Answer A is correct.

Answers B, C, and D are all valid causes for persistent oil leaks from a final drive assembly. Poor quality gaskets and sealants are the LEAST-Likely cause of these leaks.

Question #40

Answer A is correct because a smooth dull surface texture change is a clear sign of mating surface wear. This surface texture is caused by movement of the two housings eventually wearing enough to cause mating problems.

Answer B is wrong because gouges or abrupt markings are signs of assembly or disassembly damage, not wear.

Answer C is wrong because imperfections in the surface are machining marks, not signs of wear.

Answer D is wrong because pitting or light surface rust is a sign of oxidation, not a sign of wear.

Question #41

Answer A is correct. Technician A is right because a pull-type clutch has the adjuster ring inside the clutch cover. To adjust the clutch, this ring must be rotated. Technician B is wrong because this type of pull-type clutch is a ceramic button or friction disc style that is found on primarily heavy-duty trucks.

Question #42

Answer B is correct. Technician A is wrong because every time you remove a hub from an oil-lubricated type axle bearing you should pre-lube the bearing and fill the hub cavity with fresh oil. Technician B is correct because every time you remove a hub from a grease-lubricated type axle bearing you should repack the bearing with grease.

Question #43

Answer A is correct because a release bearing that is not moving freely will not cause cracks on only one side of the intermediate plate. It could cause excess heat buildup on all clutch surfaces if it was binding enough to prevent full clutch engagement.

Answer B is wrong because a poorly manufactured friction disc may have a lower coefficient of friction, and slippage could occur causing cracking on the intermediate plate.

Answers C and D are wrong because a binding friction disc or intermediate plate would cause drag on one side of the intermediate plate generating heat, which could cause cracks.

Question #44

Answer C is correct. Both technicians are right. Technician A is correct because clutch teeth on a gear should have a beveled edge to allow for easy meshing with the clutch collar. Technician B is also correct because if the clutch teeth on a gear are worn, it could cause the transmission to jump out of gear.

Question #45

Answer B is correct. Technician A is wrong because the detent has no direct affect on clutch wear. Detents only hold shift rails in their selected position. Technician B is correct because a broken detent spring in the shift tower will cause the transmission to jump out of gear.

Question #46

Answers A and B are wrong because driving in rear-wheel drive mode should not have any effect on the operation of the front axle drive selection.

Answer C is also wrong because an incorrectly adjusted range shifter should only affect the movement for low to high range.

Answer D is correct because worn teeth could produce incomplete engagement or a partial lock condition.

Question #47

Answer A is correct because worn thrust washers typically cause axial end play in the side or spider gears. They are the wear item.

Answer B is wrong because the gears do not need to be replaced unless the gear teeth are worn or otherwise damaged.

Answer C is wrong because the gears do not need to be replaced, just the thrust washers.

Answer D is wrong because it is necessary to split the case for access to the side pinion gears and thrust washers.

Question #48
Answers A, B, and C are wrong because automatic transmission fluid should be checked for the correct color (not burnt, black), checked for a burnt smell, and for the presence of particles circulating in the fluid. **Answer D is correct** because checking the transmission fluid's viscosity is not performed.

Question #49
Answer B is correct. Technician A is wrong because clutch brakes are not used with push-type clutches. Technician B is right because push-type clutches are normally found on light and medium duty vehicles.

Question #50
Answer A is wrong because overfilling can cause overheating due to aeration. Air will not remove heat from components as well as oil will. Aerated oil will not lubricate as well as non-aerated oil, which increases heat due to increased friction.
Answer B is correct because excessive clutch wear does not result from overfilling. The clutch is a separate component from the transmission and only driveability problems generated from the transmission can cause clutch wear.
Answer C is wrong because overfilling can cause leakage.
Answer D is wrong because overfilling can cause wear to gears and bearings due to air in the oil, which reduces the lubricating effect of the oil.

Question #51
Answer B is correct.
Answers A, C, and D are wrong because the correct wheel-bearing endplay should fall between 0.001 and 0.005 of an inch or the entire adjustment procedure must be performed again.

Question #52
Answer B is correct. Technician A is wrong because using a hammer and a chisel is not an appropriate method of removing the rivets. This method could cause damage to the rivet holes. Technician B is correct because the recommended procedure requires a drill and punch to remove the old rivets. This method should not cause any damage to the components.

Question #53
Answer A is correct because improper driveline set-up can cause damaging vibrations and whipping, which could result in damage to the auxiliary housing.
Answer B is wrong because worn main shaft bearings should only affect main box shifting and main shaft wear.
Answer C is wrong because a defective auxiliary synchronizer will affect range shifting.
Answer D is wrong because engine to transmission misalignment will cause input shaft, input gear, and bearing wear.

Question #54
Answers A, B, and C are wrong because improperly adjusted linkage, out of adjustment hydraulic slave cylinder, or a seized pilot bearing may all cause the clutch to remain engaged. All of these selections can affect the disengagement travel available to the release bearing.
Answer D is correct because a worn clutch disc is the LEAST-Likely cause of a clutch remaining engaged.

Question #55
Answer B is correct.
Answers A, C, and D are wrong because although all of these conditions may cause damage to the yoke or the universal joint, they will not affect yoke bore misalignment. Excessive driveline torque will place extreme twisting and separation forces on the yoke bores, which can cause distortion and misalignment in the yoke.

Question #56
Answer B is correct. Technician A is wrong because the distance from the clutch mounting surface and the friction surface is predetermined to allow for the proper clamping force from the pressure plate. To grind one surface without grinding the other would change the distance, either increasing or decreasing the clamping force.

Question #57
Answer C is correct. Both technicians are right. Technician A is correct because the wear compensator is a replaceable part. Technician B is also correct because the wear compensator will keep the free travel in the clutch pedal within specifications.

Question #58
Answer A is correct because you should visually inspect the mount while it is still in the vehicle. Looking for swelling and putting force against the mount are acceptable checks.
Answer B is wrong because removal of the mount is not recommended when inspecting. Without the proper amount of weight on the mount, swelling and some cracks may not be detectable.

Question #59
Answer A is correct because the slave valve shown is a piston type.
Answer B is wrong because the poppet type slave valve is shaped differently but is quite similar in operation.
Answers C and D are both wrong because these are ports on a slave valve and not actual types.

Question #60
Answer A is wrong because it is not standard for manufacturers to provide timing marks on the gears.
Answer B is correct because you mark the gears before disassembly, then align those marks during assembly.
Answer C is wrong because this is not an appropriate way to time the gears. The position of the main shaft is critical to the timing setting.
Answer D is wrong because there is no specific tooth and no timing marks on the main shaft.

Question #61
Answer A is wrong because attaching the dial indicator base and the indicator on the flywheel face and turning the flywheel will not produce a measurement.
Answer B is correct because when checking housing bore, and rotate the flywheel. This produces an actual runout reading and not a combined reading of run out and end play.
Answer C is wrong because the flywheel should be pushed in to take the measurement.
Answer D is wrong because flywheel run out needs to be verified, not just suspected.

Question #62
Answer A is correct. Technician A is correct because end play is the axial movement, or movement along the shaft. The dial indicator must be set in a way that it can read the amount of in and out movement of the crankshaft. Technician B is wrong because this setup would measure crankshaft radial movement.

Question #63
Answer B is correct. Technician A is wrong because even if all the necessary adjustments are made properly, they will not ensure proper operation in the future. Wear in some areas may cause part failure in the near future. Technician B is correct because the shift linkage should always be checked.

Question #64
Answers A, B, and C are wrong because the differential lock cylinder requires air pressure to move the collar to the lock position and damaged shift collar teeth would also prevent locking.
Answer D is correct because the cylinder return spring is responsible for disengaging the shift collar to unlock the differential.

Question #65

Answer A is correct because a broken detent spring is the LEAST-Likely cause of a noisy transmission. This spring is not a moving part and will not make noise when defective.

Answers B, C, and D are wrong because the input bearing, output bearing, and countershaft bearings are all moving parts in the transmission. If any of these bearings are pitted or worn they can be a source of noise.

Question #66

Answer A is correct. Technician A is correct because automatic transmission gaskets cannot be replaced by silicone sealants. Possible silicone inhalation into the hydraulic system of the transmission is the reason for not using these sealants. Technician B is wrong because it is not an accepted practice to seal a porous automatic transmission case with silicone.

Question #67

Answers A and B are wrong because a high to low range shift requires air to shift the axle into high range first. A faulty air compressor or an air leak would prevent this.

Answer C is correct because the shift to low range could not take place if the air could not exhaust through the quick release valve.

Answer D is also wrong because the plugged air filter could cause a slow air buildup and high temperature air, which will usually not affect the shift from high to low range.

Question #68

Answer A is wrong because oil contamination may cause slippage and disc failure.

Answer B is wrong because worn torsion springs may cause clutch disc hub damage.

Answer C is correct because worn U-joints will not cause premature clutch failure. This may cause driveline noise and vibration.

Answer D is wrong because worn clutch linkage may cause disc failure due to incomplete clutch engagement or disengagement.

Question #69

Answer B is correct. Technician A is wrong because the air-operated clutch systems air supply travels through a one way check valve to prevent leakage back to the supply system. Technician B is right because this system has its own isolated supply reservoir that should maintain enough air pressure for clutch application during startup even when the truck's supply system is drained down.

Question #70

Answer C is correct. Technician A is right because a sensor that is a greater distance than specified from the reluctor wheel will produce less output voltage. Technician B is also right because wheel speed sensors generate AC current. All inductive or magnetic pickup type sensors produce alternating current. Both technicians are right.

Question #71

Answer A is correct. Technician A is right because it is a recommended practice to remove the bearings and inspect them for damage and wear. Technician B is wrong because oil coating the bearings and pouring oil into the hub are good practices, but the wheel hubs must be filled to the correct level before operating the truck. Raising the opposite side of the axle and letting lubricant flow into the serviced bearing hub and topping up the axle afterwards performs this.

Question #72

Answer B is correct. Technician A is wrong because a fault code for a defective tailshaft speed sensor could be set by any fault in that circuit. Circuit wiring or a connector problem could also set this code. Technician B is correct because a multimeter can check the circuit and sensor integrity as well as sensor input and output.

Question #73
Answer A is wrong because tightening the end cap adjuster rings would indicate adjustment not inspection of bearing wear.
Answer B is wrong because tightening the end caps is not an accurate way to duplicate noise. Bearings usually generate noise under operating loads.
Answer C is correct because the figure shows someone trying to adjust differential side bearing play.
Answer D is wrong because adjusting differential side bearing play is the correct procedure and will not cause any damage.

Question #74
Answers A, B, and C are wrong because both the control valve and slave valve will operate correctly. Only the air sent from the slave valve will be sent to the wrong side of the range cylinder.
Answer D is correct because the air lines to the range cylinder are crossed. The high range feed line is connected to the low side of the range cylinder, which will cause a low range shift when high range is selected.

Question #75
Answer D is correct. Technician A is wrong because ring gear run out is measured with the side bearings preloaded to prevent differential case movement. Technician B is also wrong because preloading the bearings will not correct ring gear run out. Both technicians are wrong.

Question #76
Answer C is correct. Technician A is right because the thrust screw and thrust block should be turned inward until the thrust block contacts the ring gear and, backed then off ½ turn. Technician B is also right because under heavy loads the separating forces between the pinion and ring gear will produce enough thrust to make the ring gear contact the thrust block. This is the purpose of the thrust block to prevent ring gear deflection. Both technicians are right.

Question #77
Answer C is correct. Both technicians are correct. The high torque loads on the reverse idler gears during reverse operation, and the separating forces, due to not being placed between two gears, promotes wear on the idler shafts.

Question #78
Answers A, B, and D are wrong because these are all valid checks. Worn or oblong detent recesses may allow the detent balls to wedge in the rail or housing due to the excess wear. Broken detent springs could prevent the detent balls from seating firmly in the recesses. A rough or worn detent ball could wedge in a recess and lock the shift rail in the selected position.
Answer C is correct because detent springs do not require lubrication.

Question #79
Answer A is correct. Overtightening of the transmission can cause undue pressure on the housing face. Swollen bolt holes from overtightened bolts, and housing distortion is a likely cause of irregularities on the housing face.
Answer B is wrong because an overheated clutch disc and pressure plate is not a likely cause of housing face run out but it can cause clutch failure.
Answer C is wrong because it would be unlikely that excessive flywheel surface run out would have any effect on the clutch housing. Clutch operation would be affected.
Answer D is wrong because manufacturing defects is also an unlikely cause.

Question #80
Answers A, B, and C are wrong because rough, binding actions and spalling are all valid reasons for pilot bearing replacement. All of these conditions suggest deterioration of the bearing or races.
Answer D is correct because excessive end play is the exception. A pilot bearing is not designed to control end play.

Question #81
Answer A is correct. Technician A is correct because a clutch brake needs to be inspected for wear and fatigue in the same manner as the pressure plate. The friction surface can become contaminated from oil and it can wear due to normal and incorrect operation. Technician B is wrong because the friction face needs to be inspected.

Question #82
Answer C is correct. Technician A is correct because both the planetary double reduction and the locking differential use an air cylinder and shift fork. The cylinders are air applied and spring released. Technician B is also correct because one type of double reduction axle uses two helically cut gears as the second reduction. The pinion drives a ring that is mounted on a separate shaft along with a helical gear. This helical gear drives another larger helical gear that is mounted on the differential case.

Question #83
Answers A, C, and D are wrong because these are all valid items to inspect when replacing U-joints on a driveshaft.
Answer B is correct because final drive operating angle measurement is performed only when vibrations in the driveline are present, excess transmission output shaft radial play, loose suspension components are found or irregular tread wear is found on the rear tires.

Question #84
Answer A is correct because the engine is not able to propel both the front and rear drive axles. The input shaft of the interaxle differential drives the differential pinion gears. If the pinion gears have failed, no drive will be available to either side gears that drive the axles. This makes answers B and C incorrect. Answer D is also wrong because a completely failed interaxle differential would not have an effect on shifting between high and low speed but the axles would still be inoperative.

Question #85
Answer A is wrong because you do not cool the ring gear in a freezer overnight. This would contract the ring gear.
Answer B is correct because heating the ring gear will expand it allowing it to fit over the flywheel. Answer C is wrong because heating the flywheel and cooling the ring gear would make the flywheel expand and the ring gear contract, which would make installation impossible.
Answer D is wrong because the flywheel should not be cooled.

Question #86
Answer C is correct. Both technicians are correct. Technician A says that clearance in the U-joint will allow movement and low speed can allow the shaft to work back and forth in the joint creating a clunking noise. Technician B is also correct because a dry, underlubricated joint can also generate similar noises as the rollers bind against the trunnions and bearing cups.

Question #87
Answer C is correct. Both technicians are correct because transmission mounts absorb torque from the engine and driveline vibrations. These mounts provide a cushioned, non-rigid mounting for the transmission, which allows the transmission to wind up under engine load and absorb shock from drivetrain vibrations.

Question #88
Answer B is correct because the output bearing is at the rear of the main shaft and not directly connected to the input shaft.
Answers A, C, and D are wrong because the front bearing retainer, pilot bearing, and input bearing are all directly connected to the input shaft and should be inspected at the same time.

Question #89
Answer A is correct because the tooth contact pattern is too close to the root of the ring gear teeth. By moving the pinion away from the ring gear the pattern will move toward the pitch line of the tooth. Movement of the ring gear will move the pattern between the toe and the heel.

Question #90
Answers A, B, and C are wrong because the axle housing mounting flange and the carrier mounting flange must be free of nicks, scratches, and burrs and the axle housing bolt holes or studs must be in good condition before installing a final drive carrier.
Answer D is correct because ring gear run out should have been checked during final drive reassembly.

Question #91
Answer C is correct because the correct tool and process for yoke removal is shown in the figure. Answers A, B, and D are wrong because the tool used is a yoke puller and it cannot be used for straightening yokes or pressing seals into place. This tool cannot install a yoke because output shafts and drive pinions do not have threaded holes in their ends to allow for reverse operation of this tool.

Question #92
Answer A is correct because you do press out the old one, heat the new ring gear in oil or water, and reassemble.
Answer B is wrong because the ring gear is a interference fit component; therefore, force will be necessary.
Answer C is wrong because it is not correct to press a new ring gear onto the differential case; damage could occur while pressing the ring gear.
Answer D is wrong because it is not acceptable to use a hammer to remove the old ring gear, and heating the differential case will expand it and make it impossible to successfully mount the new ring gear.

Question #93
Answers A and B are wrong because the correct gear contact pattern will not allow any tooth meshing past the surface of the gear teeth.
Answer C is wrong because the correct pattern comes close to, but does not touch, any edge of the gear tooth. This would include a pattern that is close to the heel, which is incorrect. A correct pattern does follow these rules but specifies closeness to the toe.
Answer D is correct because the correct gear contact pattern should come as close to the edge of the toe as possible without clearing the tooth.

Question #94
Answer A is wrong because a faulty interlock mechanism will only affect shifting in the main box.
Answer B is correct because improper driveline angularity will create speed fluctuations and harmful vibrations that affect the output shaft, bearings, and synchronizers.
Answer C is wrong because a twisted mainshaft will affect shifting of the shift collars and gear timing.
Answer D is wrong because improper towing can affect mainshaft bearings and thrust washers.

Question #95
Answers A and C are wrong because replacing the internal and external hose or lines is not necessary unless they are damaged or worn.
Answer B is wrong because the pump does not need to be packed because it is submerged.
Answer D is correct because you do check the pump for smooth operation and blow forced air through the passages to remove any dirt or foreign particles that may be present.

Question #96
Answer A is wrong because the air shift system is only used for the range and splitter operation.
Answer B is correct because the shift collar can bind as it travels over the twisted section of the mainshaft.
Answer C is wrong because a worn synchronizer will affect the timing of the gear selection but not on the release of the gear.
Answer D is wrong because the clutch brake should only be used on initial engagement.

Question #97

Answer A is wrong because using the same shim size would not account for the slight amount of increased bearing size caused by pressing the bearing onto the shaft.

Answer B is wrong because a smaller shim would not set the correct amount of preload.

Answer C is correct because you note the shim size used and choose one that is 0.001 inch (0.025 mm) larger for installation which compensates for slight bearing growth during installation.

Answer D is wrong because it would be acceptable to use the shim except that it needs to be 0.001 inch (0.025 mm) larger.

Question #98

Answer A is wrong because loosening the adjusting ring would introduce unwanted free play in the bearings.

Answer B is wrong because the preload is not yet set correctly.

Answer C is correct because you do tighten each bearing adjusting ring one notch to set the preload.

Answer D is wrong because this is not necessary because the preload reading accounts for both bearings.

Question #99

Answer B is correct. Technician A is wrong because the pilot bearing does require regular maintenance. Technician B is correct because the bell housing misalignment would force the transmission input shaft to operate at a different angle than the pilot bearing, which will cause binding and undue stress on the races of the pilot bearing and eventual failure.

Question #100

Answer A is correct because a broken shift spring will cause the tractor to not shift out of interlock. The interlock mechanism is air applied and spring pressure released.

Answers B, C, and D are wrong because a broken shift shaft, an air line problem, or a stripped clutch collar would not allow the unit to shift into differential lock.

Question #101

Answer A is wrong because a crimp splice connector is not an appropriate way to fix the fault or any other wire repair.

Answer B is correct because you must replace the entire wheel speed sensor to fix the vehicle. The sensor leads are shielded and must have proper insulation to prevent interference.

Question #102

Answers A and B are wrong because a worn countershaft gear or bearing would make noise whenever rotated, which is whenever the input shaft is turning.

Answer C is wrong because a worn rear main shaft support bearing would make a noise whenever the main shaft turns, which is in every gear.

Answer D is correct because the front main shaft support bearing will only make noise when there is a speed difference between the main shaft and input shaft, which is in every gear but direct.

Question #103

Answer A is correct. Technician A is correct because scan tools can obtain transmission identification, trouble codes, perform operational tests, and download data. Technician B is wrong because the gear ratios are obtained by the mechanical gearing in the transmission. Gears must be physically changed to change a ratio.

Question #104

Answer A, B, and C are wrong because for a pilot bearing to make a noise, there must be a speed difference between the input shaft and the flywheel and all of these answers do not produce this.

Answer D is correct because sitting at a red light with the truck in gear has the input shaft stationary and the flywheel rotating.

Glossary

Actuator A device that delivers motion in response to an electrical signal.

Adapter The welds under a spring seat to increase the mounting height or fit a seal to the axle.

Adapter Ring Component bolted between the clutch cover and the flywheel on some two-plate clutches when the clutch is installed on a flat flywheel.

A/D Converter Abbreviation for Analog-to-Digital Converter.

Additive An additional element intended to improve a certain characteristic of the material.

Adjusting Ring A device that is held in the shift signal valve bore by a press fit pin through the valve body housing. When the ring is pushed in by the adjusting tool, the slots on the ring that engage the pin are released.

Air Compressor An engine-driven pump for supplying compressed air to the truck brake and air system.

Air Dryer A unit that removes moisture.

Air Filter/Regulator Assembly A device that minimizes the possibility of moisture-laden air or impurities entering a system.

Air Shifting The process that uses air pressure to engage different range combinations in the transmission's auxiliary section without a mechanical linkage to the driver.

Ambient Temperature Temperature of the surrounding or prevailing air. Normally, it is considered to be the temperature in the service area where testing is taking place.

Amboid Gear A gear that is similar to the hypoid type with one exception: the axis of the drive pinion gear is located above the centerline axis of the ring gear.

Amp Abbreviation for ampere.

Ampere The unit for measuring electrical current.

Analog Signal A voltage signal that varies within a given range (from high to low, including all points in between).

Analog-to-Digital Converter (A/D converter) A device that converts analog voltage signals to a digital format; this is located in a section of the processor called the input signal conditioner.

Analog Volt/Ohmmeter (AVOM) A test meter used for checking voltage and resistance. Analog meters should not be used on solid state circuits.

Annulus Ring gear: internally toothed ring gear in a planetary gear set.

Anticorrosion A chemical used to protect metal surfaces from corrosion.

Antirattle Springs Springs that reduce wear between the intermediate plate and the drive pin, and help to improve clutch release.

Antirust Agent Additive used with lubricating oils to prevent rusting.

ASE Abbreviation for Automotive Service Excellence, a trademark of National Institute for Automotive Service Excellence.

ATEC System Electronically controlled, automatic transmission system that includes an electronic control system, torque converter, lockup clutch, and planetary gear train. Now known as CEC.

Atmospheric Pressure The weight of the air at sea level; 14.696 pounds per square inch (psi) or 101.33 kilopascals (kPa).

Autoshift Finger Device that engages the shift blocks on the yoke bars that corresponds to the tab on the end of the gearshift lever in manual systems.

Auxiliary Filter A device installed in the oil return line between the oil cooler and the transmission to prevent debris from being flushed into the transmission causing a failure. An auxiliary filter must be installed before the vehicle is placed back in service.

Auxiliary Section The section of a transmission where range shifting occurs, housing the auxiliary drive gear, auxiliary main shaft assembly, auxiliary countershaft, and the synchronizer assembly.

Axis of Rotation The center line around which a gear or part revolves.

Axle (1) A rod or bar on which wheels turn. (2) A shaft that transmits driving torque to the wheels.

Axle Range Interlock A feature designed to prevent axle shifting when the interaxle differential is locked out, or when lockout is engaged. The basic shift system operates the same as the standard shift system to shift the axle and engage or disengage the lockout.

Axle Seat A suspension component used to support and locate the spring on an axle.

Axle Shims Thin wedges that may be installed under the leaf springs of single axle vehicles to tilt the axle and correct the U-joint operating angles. Wedges are available in a range of sizes to change pinion angles.

Backing Plate A metal plate that serves as the foundation for the brake shoes and other drum brake hardware.

Bellows A movable cover or seal that is pleated or folded like an accordion to allow for expansion and contraction.

Block Diagnosis Chart A troubleshooting chart that lists symptoms, possible causes, and probable remedies in columns.

Boss A heavy cast section that is used for support, such as the outer race of a bearing.

Bottoming A condition that occurs when: (1) The teeth of one gear touch the lowest point between teeth of a mating gear. (2) The bed or frame of the vehicle strikes the axle, such as may be the case of overloading.

Bottom U-Bolt Plate A plate that is located on the bottom side of the spring or axle and is held in place when the U-bolts are tightened to the clamp spring and axle together.

Bracket An attachment used to secure components to the body or frame.

Brake Disc A steel disc used in a braking system with a caliper and pads. When the brakes are applied, the pad on each side of the spinning disc is forced against the disc, thus imparting a braking force. This type of brake is very resistant to brake fade.

Brake Drum A cast metal bell-like cylinder attached to the wheel that is used to house the brake shoes and provide a friction surface for stopping a vehicle.

British Thermal Unit (Btu) A measure of heat quantity equal to the amount of heat required to raise 1 pound of water 1° F.

Broken Back Driveshaft A term often used for non-parallel driveshaft.

Btu Abbreviation for British Thermal Unit.

CEC Electronically controlled automatic transmission system manufactured by Allison.

Center of Gravity The point around which the weight of a truck is evenly distributed; the point of balance.

Check Valve A valve that allows air or fluid to flow in one direction only. It is a federal requirement to have a check valve between the wet and dry air tanks.

Circuit The complete path of an electrical current, including the power source. When the path is unbroken, the circuit is closed and current flows. When circuit continuity is broken, the circuit is open and current flow stops.

Climbing A gear problem caused by excessive wear in gears, bearings, and shafts whereby the gears move sufficiently apart to cause the apex (or point) of the teeth on one gear to climb over the apex of the teeth on another gear with which it is meshed.

Clutch A device for connecting and disconnecting the engine from the transmission; a means of coupling and uncoupling components.

Clutch Brake A circular disc with a friction surface that is mounted on the transmission input spline between the release bearing and the transmission. Its purpose is to slow or stop the transmission input shaft rotation in order to allow gears to be engaged without clashing or grinding.

Clutch Housing A component that surrounds and protects the clutch and connects the transmission case to the vehicle's engine.

Clutch Pack An assembly of normal clutch plates, friction discs, and one very thick plate known as the pressure plate. The pressure plate has tabs around the outside diameter to mate with the channel in the clutch drum.

COE Abbreviation for cab-over-engine.

Coefficient of Friction A measurement of the amount of friction developed between two objects in physical contact when one is drawn across the other.

Combination A truck coupled to one or more trailers.

Compression Applying pressure to a spring or any springy substance, thus causing it to reduce its length in the direction of the compressing force.

Compressor Mechanical device that increases pressure within a container by pumping air into it.

Condensation The process by which gas (or vapor) changes to a liquid.

Conductor Any material that permits electrical current flow.

Controlled Traction A type of differential that uses a friction plate assembly to transfer drive torque from a slipping wheel to the one wheel that has good traction or surface bite.

Coupling Point The point at which the turbine is turning at the same speed as the impeller.

Cross Groove Joint Joint disc-shaped type of inner CV joint that uses balls and V-shaped grooves on the inner and outer races to

accommodate the plunging motion of the half-shaft. The joint usually bolts to a transaxle stub flange; same as disc-type joint.

Dampen Slow or reduce oscillations or movement.

Dampened Discs Discs that have dampening springs incorporated into the disc hub. When engine torque is first transmitted to the disc, the plate rotates on the hub, compressing the springs. This action absorbs the shocks and torsional vibration caused by today's low rpm, high torque, engines.

Dash Control Valves A variety of hand-operated valves located on the dash. They include parking-brake valves, tractor-protection valves, and differential lock.

Data Links Circuits through which computers communicate with other electronic devices such as control panels, modules, some sensors, or other computers in the form of digital signals.

Dead Axle Non-live or dead axles are often mounted in lift suspensions. They hold the axle off the road when the vehicle is traveling empty, and put it on the road when a load is being carried.

Deadline To take a vehicle out of service.

Deburring To remove sharp edges from a cut.

Deflection Bending or moving to a new position as the result of an external force.

Department of Transportation (DOT) A government agency that establishes vehicle standards.

Detergent Additive An additive that helps keep metal surfaces clean and prevents deposits. These additives suspend particles of carbon and oxidized oil in the oil.

Diagnostic Flow Chart A chart that provides a systematic approach to component troubleshooting and repair. They are found in service manuals and are vehicle make and model specific.

Dial Caliper Measuring instrument capable of taking inside, outside, depth, and step measurements.

Differential A gear assembly that transmits torque from the driveshaft to the wheels and allows two opposite wheels to turn at different speeds for cornering and traction.

Differential Carrier Assembly An assembly that controls the drive axle operation.

Differential Lock A toggle or push-pull type air switch that locks together the rear axles of a tractor so they pull as one for off-the-road operation.

Digital Binary Signal A signal that has only two values; on and off.

Digital Volt/Ohmmeter (DVOM) Test meter recommended by most manufacturers for use on solid state circuits.

Diode The simplest semiconductor device formed by joining P-type semiconductor material with N-type semiconductor material. A diode allows current to flow in one direction, but not in the opposite direction.

Direct Drive The gearing of a transmission so that one revolution of the engine produces one revolution of the transmission's output shaft. The drive ratio of a direct drive transmission would be 1:1.

Dog Tracking Off-center tracking of the rear wheels as related to the front wheels.

DOT Abbreviation for Department of Transportation.

Downshift Control The selection of a lower range to match driving conditions encountered or expected to be encountered. Learning to take advantage of a downshift gives better control on slick or icy roads and on steep downgrades. Downshifting to lower ranges increases engine braking.

Double Reduction Axle An axle that uses two gear sets for greater overall gear reduction and peak torque development. This design is favored for severe service applications, such as dump trucks, cement mixers, and other heavy haulers.

Driven Gear A gear that is driven or forced to turn by a drive gear, by a shaft, or by some other device.

Drive or Driving Gear A gear that drives another gear or causes another gear to turn.

Driveline The propeller or driveshaft and universal joints that transmit transmission output to the axle pinion gear shaft.

Driveline Angle The alignment of the transmission output shaft, driveshaft, and rear axle pinion centerline.

Driveshaft An assembly of one or two universal joints connected to a shaft or tube; used to transmit power from the transmission to the differential.

Drivetrain An assembly that includes all power transmitting components from the engine to the wheels, including clutch/torque converter, transmission, driveline, and front and rear driving axles.

Driver Controlled Main Differential Lock A type of axle assembly that has greater flexibility over the standard type of single reduction axle because it provides equal amounts of driveline torque to each driving wheel, regardless of changing road conditions. This design also provides the necessary differential action to the road wheels when the truck is turning a corner.

Driver's Manual A publication that contains information needed by the driver to understand, operate, and care for the vehicle and its components.

Drum Brake A type of brake system in which stopping friction is created by the shoes pressing against the inside of the rotating drum.

ECU Abbreviation for electronic control unit.

Electricity The movement of electrons from one place to another.

Electric Retarder Electromagnets mounted in a steel frame. Energizing the retarder causes the electromagnets to exert a dragging force on the rotors in the frame and this drag force is transmitted directly to the driveshaft.

Electromotive Force (EMF) The force that moves electrons between atoms. This force is the pressure that exists between the positive and negative points (the electrical imbalance). This force is measured in units called volts.

Electronically Programmable Memory (EPROM) Computer memory that permits adaptation of the ECU to various standard mechanically controlled functions.

Electronic Control Unit (ECU) System control module.

Electronics The technology of controlling electricity.

Electrons Negatively charged particles orbiting around every nucleus.

EMF Abbreviation for electromotive force. AKA: voltage.

End Yoke The component connected to the output shaft of the transmission to transfer engine torque to the driveshaft.

Engine Brake A hydraulically operated device that converts the vehicle engine into a power absorbing retarding mechanism.

Engine Stall Point The rpm, under load specified for the stall test.

EPROM Abbreviation for Electronically Programmable Memory.

External Housing Damper A counterweight attached to an arm on the rear of the transmission extension housing and designed to dampen unwanted driveline or powertrain vibrations.

False Brinelling The polishing of a surface that is not damaged.

Fatigue Failures The progressive destruction of a shaft or gear teeth material usually caused by overloading.

Fault Code A code that is recorded into the computer's memory. A fault code can be read by connecting an EST (electronic service tool) into the computer.

Federal Motor Vehicle Safety Standard (FMVSS) A federal standard that specifies that all vehicles in the United States be assigned a Vehicle Identification Number (VIN).

Final Drive The last reduction gear set of a truck.

Fixed Value Resistor Electrical device designed to have only one resistance rating, which should not change, for controlling voltage.

Flammable Any material that will easily catch fire or explode.

Flare To spread gradually outward in a bell shape.

Flex Disc Term often used for flex plate.

Flexplate Component used to mount the torque converter to the crankshaft. The flexplate is positioned between the engine crankshaft and the T/C. The purpose of the flexplate is to transfer crankshaft rotation to the shell of the torque converter assembly.

Float A cruising drive mode in which the throttle setting matches engine speed to road speed, neither accelerating nor decelerating.

Floating Main Shaft The main shaft consisting of a heavy-duty central shaft and several gears that turn freely when not engaged. The main shaft can move to allow for equalization of the loading on the countershafts. This is key to making a twin countershaft transmission workable. When engaged, the floating main shaft transfers torque evenly through its gears to the rest of the transmission and ultimately to the rear axle.

FMVSS Abbreviation for Federal Motor Vehicle Safety Standard.

Foot-Pound English unit of measurement for torque. One foot-pound is the torque obtained by a force of 1 pound applied to a foot long wrench handle.

Forged Journal Cross Part of a universal joint. Trunnion.

Fretting A result of vibration that the bearing outer race can pick up the machining pattern.

Friction Plate Assembly An assembly consisting of a multiple disc clutch that is designed to slip when a predetermined torque value is reached.

Fully Floating Axles An axle configuration whereby the axle half shafts transmit only driving torque to the wheels and not bending and torsional loads that are characteristic of the semi-floating axle.

Fusible Link A term often used for fuse link.

Fuse Link A short length of smaller gauge wire installed in a conductor, usually close to the power source.

GCW Abbreviation for gross combination weight.

Gear A disc-like wheel with external or internal teeth that serves to transmit or change motion.

Gear Pitch The number of teeth per given unit of pitch diameter, an important factor in gear design and operation.

Gladhand The connectors between tractor and trailer air lines.

Gross Combination Weight (GCW) The total weight of a fully quipped vehicle including payload, fuel, and driver.

Gross Vehicle Weight (GVW) The total weight of a fully equipped vehicle and its payload.

Ground The negatively charged side of a circuit. A ground can be a wire, the negative side of the battery, or the vehicle chassis.

Grounded Circuit A shorted circuit that causes current to return to the battery before it has reached its intended destination.

GVW Abbreviation for gross vehicle weight.

Harness and Harness Connectors The organization of the vehicle's electrical system providing an orderly and convenient starting point for tracking and testing circuits.

Hazardous Materials Any substance that is flammable, explosive, or is known to produce adverse health effects in people or the environment that are exposed to the material during its use.

Heads Up Display (HUD) A technology used in some vehicles that superimposes data on the driver's normal field of vision. The operator can view the information, which appears to "float" just above the hood at a range near the front of a conventional tractor or truck. This allows the driver to monitor conditions such as limited road speed without interrupting his normal view of traffic.

Heat Exchanger A device used to transfer heat, such as a radiator or condenser.

Heavy-Duty Truck A truck that has a GVW of 26,001 pounds or more.

High-Resistant Circuits Circuits that have an increase in circuit resistance, with a corresponding decrease in current.

Hinged Pawl Switch The simplest type of switch; one that makes or breaks the current of a single conductor.

HUD Abbreviation for heads up display.

Hypoid Gears A bevel gear crown and pinion assembly where the axes are at right angles but the pinion is on a lower plane than the crown. Hypoid gearing uses a modified spiral bevel gear structure that allows several gear teeth to absorb the driving power and allows the gears to run quietly. A hypoid gear is typically found at the drive pinion gear and ring gear interface.

Inboard Toward the centerline of the vehicle.

In-Line Fuse A fuse that is in series with the circuit in a small plastic fuse holder, not in the fuse box or panel. It is used, when necessary, as a protection device for a portion of the circuit even though the entire circuit may be protected by a fuse in the fuse box or panel.

In-Phase The in-line relationship between the forward coupling shaft yoke and the driveshaft slip yoke of a two-piece driveline.

Input Retarder A device located between the torque converter housing and the main housing designed primarily for over-the-road operations. The device employs a "paddle wheel" type design with a vaned rotor mounted between stator vanes in the retarder housing.

Installation Templates Drawings supplied by some vehicle manufacturers to allow the technician to correctly install the accessory. The templates available can be used to check clearances or to ease installation.

Insulator A material, such as rubber or glass, that offers high resistance to the flow of electrons.

Integrated Circuit A solid state component containing diodes, transistors, resistors, capacitors, and other electronic components.

Jumper Wire A wire used to temporarily bypass a circuit or components for electrical testing. A jumper wire consists of a length of wire with an alligator clip at each end.

Jump Out A condition that occurs when a fully engaged gear and sliding clutch are forced out of engagement.

Kinetic Energy Energy in motion.

Limited-Slip Differential A differential that uses a clutch device to deliver torque to either rear wheel when the opposite wheel is spinning.

Linkage A system of rods and levers used to transmit motion or force.

Live Axle An axle on which the wheels are firmly affixed. The axle drives the wheels.

Live Beam Axle A non-independent suspension in which the axle moves with the wheels.

Lockstrap A manual adjustment mechanism that allows for the adjustment of free travel.

Lockup Torque Converter A torque converter that eliminates the 10 percent slip that takes place between the impeller and turbine at the coupling stage of operation. It is considered a four-element (impeller, turbine, stator, lockup clutch), three-stage (stall, coupling, and locking stage) unit.

Maintenance Manual A publication containing routine maintenance procedures and intervals for vehicle components and systems.

Main Transmission A transmission consisting of an input shaft, floating main shaft assembly and main drive gears, two countershaft assemblies, and reverse idler gears.

Multiple Disc Clutch A clutch having a large drum-shaped housing that can be either a separate casting or part of the existing transmission housing.

NATEF Abbreviation for National Automotive Education Foundation.

National Automotive Technicians Education Foundation (NATEF) A foundation having a program of certifying secondary and post secondary automotive and heavy-duty truck training programs.

National Institute for Automotive Service Excellence (ASE) A nonprofit organization that has an established certification program for automotive, heavy-duty truck, auto body repair, engine machine shop technicians, and parts specialists.

Needlenose Pliers This tool has long tapered jaws for grasping small parts or for reaching into tight spots. Many needlenose pliers also have cutting edges and a wire stripper.

NIASE Abbreviation for National Institute for Automotive Service Excellence, now abbreviated ASE.

NIOSH Abbreviation for National Institute for Occupation Safety and Health.

NLGI Abbreviation for National Lubricating Grease Institute.

NHTSA Abbreviation for National Highway Traffic Safety Administration.

Non-live Axle Non-live or dead axles are often mounted in lift suspensions. They hold the axle off the road when the vehicle is traveling empty, and put it on the road when a load is being carried.

Nonparallel Driveshaft A type of driveshaft installation whereby the working angles of the joints of a given shaft are equal; however the companion flanges and/or yokes are not parallel.

OEM Abbreviation for original equipment manufacturer.

Off-road With reference to unpaved, rough, or ungraded terrain on which a vehicle will operate. Any terrain not considered part of the highway system falls into this category.

Ohm A unit of measured electrical resistance.

Ohm's Law The basic law of electricity stating that in any electrical circuit, current, resistance, and pressure work together in a mathematical relationship.

On-road With reference to paved or smooth-graded surface terrain on which a vehicle will operate, generally considered to be part of the public highway system.

Open Circuit An electrical circuit whose path has been interrupted or broken either accidentally (a broken wire) or intentionally (a switch turned off).

Oscillation Rotational movement in either fore/aft or side-to-side direction about a pivot point.

OSHA Abbreviation for Occupational Safety and Health Administration.

Out-of-Phase A condition of the universal joint which acts somewhat like one person snapping a rope held by a person at the opposite end. The result is a counter reaction at the opposite end. If both were to snap the rope at the same time, the resulting waves cancel each other and neither would feel the reaction.

Output Driver An electronic on/off switch that the computer uses to control the ground circuit of a specific actuator. Output drivers are located in the processor along with the input conditioners, microprocessor, and memory.

Output Yoke Component that serves as a connecting link, transferring torque from the transmission's output shaft through the vehicle's driveline to the rear axle.

Oval A condition that occurs when a tube or bore is not round: eccentric.

Overall Ratio The ratio of the lowest to the highest forward gear in the transmission.

Overdrive The gearing of a transmission so that in its highest gear one revolution of the engine produces more than one revolution of the transmission's output shaft.

Overrunning Clutch A clutch mechanism that transmits torque in one direction only.

Oxidation Inhibitor An additive used with lubricating oils to keep oil from oxidizing even at very high temperatures.

Parallel Circuit An electrical circuit that provides two or more paths for the current to flow.

Parallel Joint Type A type of driveshaft installation whereby all companion flanges and/or yokes in the driveline are parallel to each other with the working angles of the joints of a given shaft being equal and opposite.

Parking Brake A mechanically applied brake used to prevent a parked vehicle's movement.

Parts Requisition A form that is used to order new parts, on which the technician writes the names of what part(s) are needed along with the vehicle's VIN or company's identification folder.

Payload The weight of the cargo carried by a truck, not including the weight of the chassis.

Pitting Surface irregularities resulting from corrosion.

Planetary Drive A planetary gear reduction set where the sun gear is the drive and the planetary carrier is the output.

Planetary Gear Set A system of gearing that is somewhat like the solar system. A pinion is surrounded by an internal ring gear and planet gears are in mesh between the ring gear and pinion around which all revolve.

Planetary Pinion Gears Small gears fitted into a framework called the planetary carrier.

Polarity The state, either positive or negative, with reference to the two poles or to electrification.

Pole The number of input circuits made by an electrical switch.

Pounds per Square Inch (psi) A unit of English measurement for pressure.

Power A measure of work being done factored with time.

Power Flow The flow of power from the input shaft through one or more sets of gears to the transmission output shaft.

Power Synchronizer A device to speed up the rotation of the main section gearing for smoother automatic downshifts and to slow down the rotation of the main section gearing for smoother automatic upshifts.

Powertrain An assembly consisting of a driveshaft, coupling, clutch, and transmission differential.

Pressure The amount of force applied to a definite area measured in pounds per square inch (psi) English or kilopascals (kPa) metric.

Pressure Differential The difference in pressure between any two points of a system or a component.

Printed Circuit Board An electronic circuit board made of thin nonconductive plastic-like material onto which conductive metal, such as copper, has been deposited. Parts of the metal are then etched away by an acid, leaving metal lines that form the conductors for the various circuits on the board.

Programmable Read-Only Memory (PROM) An electronic component that contains program information specific to different vehicle model calibrations.

PROM Abbreviation for Programmable Read-Only Memory.

Priority Valve A valve that ensures that the control system upstream from the valve will have sufficient pressure during shifts to perform its automatic functions.

psi Abbreviation for pounds per square inch.

Pull-Type Clutch A type of clutch that does not push the release bearing toward the engine; instead, it pulls the release bearing toward the transmission.

Pump/Impeller Assembly The input (drive) member that receives torque from the engine.

Push Circuit A circuit that raises the cab from the lowered position to the desired tilt position.

Push-Type Clutch A type of clutch in which the release bearing is not attached to the clutch cover.

Radial Load A load that is applied at 90° to an axis of rotation.

RAM Abbreviation for random access memory.

Random Access Memory (RAM) The memory used during computer operation to store temporary information. The microcomputer can write, read, and erase information from RAM in any order, which is why it is called random.

Range Shift Cylinder A component located in the auxiliary section of the transmission. This component, when directed by air pressure via low and high ports, shifts between high and low range of gears.

Range Shift Lever A lever located on the shift knob allows the driver to select low- or high-gear range.

Rated Capacity The maximum, recommended safe load that can be sustained by a component or an assembly without permanent damage.

Read Only Memory (ROM) A type of memory used in microcomputers to store information permanently.

Recall Bulletin A bulletin that pertains to special situations that involve service work or replacement of components in connection with a recall notice.

Reference Voltage The voltage supplied to a sensor by the computer, which acts as a base line voltage; modified by the sensor to act as an input signal.

Relay An electric switch that allows a small current to control a much larger one. It consists of a control circuit and a power circuit.

Release Bearing A unit within the clutch consisting of bearings that mount on the transmission input shaft but do not rotate with it.

Resistance The opposition to current flow in an electrical circuit.

Revolutions per Minute (rpm) The number of complete turns a member makes in one minute.

Right to Know Law A law passed by the federal government and administered by the Occupational Safety and Health Administration (OSHA) that requires any company that uses or produces hazardous chemicals or substances to inform its employees, customers, and vendors of any potential hazards that may exist in the workplace as a result of using the products.

Rigid Disc A steel plate to which friction linings, or facings, are bonded or riveted.

Rigid Torque Arm A member used to retain axle alignment and, in some cases, to control axle torque. Normally, one adjustable and one rigid arm are used per axle so the axle can be aligned.

Ring Gear (1) The gear around the edge of a flywheel. (2) A large circular gear such as that found in a final drive assembly.

Roller Clutch A clutch designed with a movable inner race, rollers, accordion (apply) springs, and outer race. Around the inside diameter of the outer race are several cam-shaped pockets. The clutch assembly rollers and accordion springs are located in these pockets.

ROM Abbreviation for read only memory.

Rotary Oil Flow A condition caused by the centrifugal force applied to the fluid as the converter rotates around its axis.

Rotation A term used to describe a device that is turning.

rpm Abbreviation for revolutions per minute.

Rotor The rotating member of an assembly, a shaft, or disc.

Run out Deviation of specified travel of an object. The deviation or wobble a shaft or wheel has as it rotates. Run out is measured with a dial indicator.

Safety Factor (SF) (1) The amount of load which can safely be absorbed by and through the vehicle chassis frame members. (2) The difference between the stated and rated limits of a product, such as a grinding disc.

Screw Pitch Gauge A gauge used to provide a quick and accurate method of checking the threads per inch of a nut or bolt.

Self-Adjusting Clutch A clutch that automatically takes up the slack between the pressure plate and clutch disc as wear occurs.

Semiconductor Solid state device that can function as either a conductor or an insulator, depending on how its structure is arranged.

Semifloating Axle An axle type in which drive torque from the differential is transferred directly to the wheels. A single bearing assembly, located at the outer end of the axle, is used to support the axle half-shaft.

Sensing Voltage The voltage that allows the regulator to sense and monitor the battery voltage level.

Sensor An electronic device used to monitor conditions for computer control requirements.

Series Circuit A circuit connected to a voltage source with only one path for electrons flow.

Series/Parallel Circuit A circuit designed so that both series and parallel combinations exist within the same circuit.

Service Bulletin A publication that provides the latest service tips, field repairs, product improvements, and related information of benefit to service personnel.

Service Manual A manual, published by the manufacturer, that contains service and repair information for all vehicle systems and components.

Shift-Bar Housing Available in standard- and forward-position configurations, a component that houses the shift rails, shift yokes, detent balls and springs, interlock balls, and pin and neutral shaft.

Shift Fork The Y-shaped component located between the gears on the main shaft that, when actuated, causes the gears to engage or disengage via sliding clutches. Shift forks are located between low and reverse, first and second, and third and fourth gears.

Shift Rail Shift rails guide the shift forks using a series of grooves, tension balls, and springs to hold the shift forks in gear. The grooves in the forks allow them to interlock the rails, and the transmission cannot be accidentally shifted into two gears at the same time.

Shift Tower The main interface between the driver and the transmission, consisting of a gearshift lever, pivot pin, spring, boot, and housing.

Shift Yoke A Y-shaped component located between the gears on the main shaft that, when actuated, causes the gears to engage or disengage via sliding clutches. Shift yokes are located between low and reverse, first and second, and third and fourth gears.

Short Circuit An undesirable electrical connection between two worn or damaged wires or ground.

Single Reduction Axle Any axle assembly that employs only one gear reduction through its differential carrier assembly.

Slave Valve A valve to help protect gears and components in the transmission's auxiliary section by permitting range shifts to occur only when the transmission's main gearbox is in neutral. Air pressure from a regulator signals the slave valve into operation.

Slipout A condition that generally occurs when pulling with full power or decelerating with the load pushing. Tapered or worn clutching teeth will try to "walk" apart as the gears rotate, causing the sliding clutch and gear to slip out of engagement.

Solenoid An electromagnet that is used to perform work, made with one or two coil windings wound around an iron tube.

Solid-State Device An electronic device that requires little power to operate, is very reliable, and generates little heat.

Solid Wire A single-strand conductor.

Solvent A substance which dissolves other substances.

Spalling Surface fatigue that occurs when chips, scales, or flakes of metal break off due to fatigue rather than wear. Spalling is usually found on splines and U-joint bearings.

Specialty Service Shop A shop that specializes in areas such as engine rebuilding, transmission/axle overhauling, brake, air conditioning/heating repairs, and electrical/electronic work.

Spiral Bevel Gear A helical gear arrangement that has a drive pinion gear that meshes with the ring gear at the centerline axis of the ring gear. This gearing provides strength and allows for quiet operation.

Splined Yoke A yoke that allows the driveshaft to increase in length to accommodate movements of the drive axles.

Spring Chair A suspension component used to support and locate the spring on an axle.

Staff Test A test performed when there is an obvious malfunction in the vehicle's power package (engine and transmission), to determine which of the components is at fault.

Stand Pipe A type of check valve which prevents reverse flow of the hot liquid lubricant generated during operation. When the universal joint is at rest, one or more of the cross ends will be up. Without the stand pipe, lubricant would flow out of the upper passageways and trunnions, leading to partially dry startup.

Starter Motor The device that converts the electrical energy from the battery into mechanical energy for cranking the engine.

Starting Safety Switch A switch that prevents vehicles with automatic transmissions from being started in gear.

Static Balance Balance at rest, or still balance. It is the equal distribution of the weight of the wheel and tire around the axis of rotation so that the wheel assembly has no tendency to rotate by itself regardless of its position.

Stator A component located between the pump/impeller and turbine to redirect the oil flow from the turbine back into the impeller in the direction of impeller rotation with minimal loss of speed or force.

Stator Assembly The reaction member or torque multiplier supported on a free wheel roller race that is splined to the valve and front support assembly.

Stepped Resistor A resistor designed to have two or more fixed values, available by connecting wires to either of the several taps.

Still Balance Balance at rest; the equal distribution of the weight of the wheel and tire around the axis of rotation so that the wheel assembly has no tendency to rotate by itself regardless of its position.

Stranded Wire Wire that is are made up of a number of small solid wires, generally twisted together, to form a single conductor.

Swage To reduce or taper.

Switch A device used to control on/off and direct the flow of current in a circuit. A switch can be under the control of the driver or can be self-operating through a condition of the circuit, the vehicle, or the environment.

Synchromesh A mechanism that equalizes the speed of the gears that are clutched together.

Synchro-transmission A transmission with mechanisms for synchronizing the gear speeds so that the gears can be shifted without clashing, thus eliminating the need for double-clutching.

Tachometer An instrument that indicates rotating speeds, sometimes used to indicate crankshaft rpm.

Tag Axle The rearmost axle of a tandem axle tractor used to increase the load-carrying capacity of the vehicle.

Tapped Resistor A resistor designed to have two or more fixed values, available by connecting wires to either of the several taps.

Tandem A pair: often used to describe drive axles on a highway tractor.

Tandem Drive A two-axle drive combination.

Tandem Drive Axle A type of axle that combines two single-axle assemblies through the use of an interaxle differential or power divider and a short shaft that connects the two axles together.

Three-Speed Differential A type of axle in a tandem two-speed axle arrangement with the capability of operating the two drive axles in different speed ranges at the same time. The third speed is actually an intermediate speed between the high and low range.

Time Guide Prepared reference material used for computing compensation payable by the truck manufacturer for repairs or service work to vehicles under warranty, or for other special conditions authorized by the company.

Timing A procedure of marking the appropriate teeth of a gear set prior to installation and placing them in proper mesh while in the transmission.

Top U-Bolt Plate A plate located on the top of the spring and held in place when the U-bolts are tightened to clamp the spring and axle together.

Torque Twisting force.

Torque Converter A device, similar to a fluid coupling, that transfers engine torque to the transmission input shaft and can multiply engine torque by having one or more stators between the members.

Torque Limiting Clutch Brake A clutch brake designed to slip when loads of 20 to 25 pound–feet (27 to 34 N) are reached protecting the brake from overloading and the resulting high heat damage.

Torque Rod Shim A thin wedge-like insert that rotates the axle pinion to change the U-joint operating angle.

Torsional Rigidity A component's ability to remain rigid when subjected to twisting forces.

Total Pedal Travel The complete distance the clutch or brake pedal must move.

Tracking The travel of the rear wheels in a parallel path with the front wheels.

Tractor A motor vehicle, without a body, that has a fifth wheel and is used for pulling a semitrailer.

Transfer Case An additional gearbox located between the main transmission and the rear axle to transfer torque from the transmission to the front and rear driving axles.

Transistor An electronic device produced by joining three sections of semiconductor materials.

Transmission A device used to transmit torque at various ratios and can also change the direction of rotation.

Transverse Vibrations A condition caused by an unbalanced driveline or bending movements in the driveshaft.

Tree Diagnosis Chart A chart used to provide a logical sequence for what should be inspected or tested when troubleshooting a repair problem.

Trunnion The end of the universal cross; they are case hardened ground surfaces on which the needle bearings ride.

TTMA Abbreviation for Truck and Trailer Manufacturers Association.

Turbine The output (driven) member that is splined to the forward clutch of the transmission and to the turbine shaft assembly.

TVW Abbreviation for (1) Total vehicle weight. (2) Towed vehicle weight.

Two-Speed Axle Assembly An axle assembly having two different output ratios from the differential. The driver selects the ratios from the controls located in the cab of the truck.

U-Bolt A fastener used to clamp the top U-bolt plate, spring, axle, and bottom U-bolt plate together. Inverted (nuts down) U-bolts cross springs when in place; conventional (nuts up) U-bolts wrap around the axle.

Universal Joint (U-joint) A component that allows torque to be transmitted to components that are operating at different angles.

Vacuum Air below atmospheric pressure.

Validity List A list supplied by the manufacturer of valid bulletins.

Valve Body and Governor Test Stand Specialized test equipment. The valve body of the transmission is removed from the vehicle and mounted into the test stand. The test stand duplicates all vehicle running conditions, so the valve body can be thoroughly tested and calibrated.

Variable Pitch Stator A stator design often used in torque converters in off-highway applications such as aggregate or dump trucks, or other specialized equipment used to transport heavy loads in rough terrain.

Vehicle Retarder Optional braking device used to assist the service brakes on heavy-duty trucks.

VIN Abbreviation for Vehicle Identification Number.

Viscosity Oil thickness or resistance to flow.

Volt Unit of electromotive force.

Voltage-Generating Sensors Devices which produce their own input voltage signal.

Voltage Limiter Device that provides protection by limiting voltage to the instrument panel gauges to approximately 5 volts.

Vortex Oil Flow The circular flow that occurs as the oil is forced from the impeller to the turbine and then back to the impeller.

Watt Measure of electrical power.

Watt's Law Law of electricity used to find the power of an electrical circuit expressed in watts. It states that power equals the voltage multiplied by the current, in amperes.

Wear Compensator Device mounted in the clutch cover having an actuator arm that fits into a hole in the release-sleeve retainer.

Wheel and Axle Speed Sensors Electromagnetic devices used to monitor vehicle speed information for an antilock controller.

Windings The coil of wire found in a relay or other similar device.

Work (1) Forcing a current through a resistance. (2) The product of a force.

Yield Strength The highest stress a material can stand without permanent deformation or damage, expressed in pounds per square inch (psi).

Yoke Sleeve Kit This can be installed instead of completely replacing a damaged yoke. The sleeve is of heavy walled construction with a hardened steel surface having an outside diameter that is the same as the original yoke diameter.

Notes

Notes

Notes

Notes

Notes

Notes

Notes

Notes

Notes

Notes

Notes

Notes

Sample Test for Practice

Sample Test

Please note the letter and number in parentheses following each question. They match the overview in Section 4 that discusses the relevant subject matter. You may want to refer to the overview using this cross-referencing key to help with questions posing problems for you.

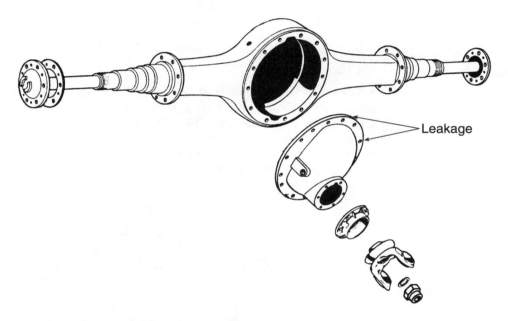

Leakage

1. Fluid leaks from the component shown in the figure are evident between the axle housing and the carrier assembly. What would the Most-Likely cause of this leak be?
   A. Damaged gasket or missing sealant.
   B. Repeated overloading of the drivetrain.
   C. Plugged axle housing breather vent.
   D. Moisture contaminated axle lubricant. (D2)

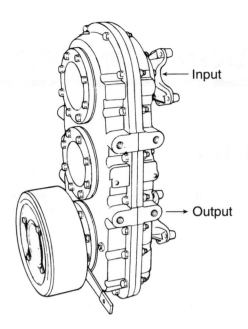

2. A truck that is equipped with a conventional three-shaft transfer case with the additional PTO and the front axle de-clutch will not drive the front axle when driving over rough terrain. The LEAST-Likely cause of the problem is:
   A. a faulty front axle de-clutch.
   B. a broken differential in the front axle.
   C. a stripped front axle ring gear.
   D. a blown fuse.   (B25)

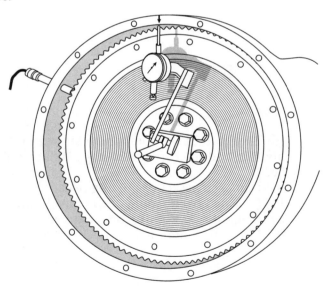

3. In the illustration above, the technician is measuring flywheel housing bore runout. The TIR or total indicated runout measurement is 0.006 inch (0.2 mm). What will the technician do next?
   A. Resurface the flywheel housing.
   B. Continue with the job.
   C. Service the flywheel.
   D. Replace the pilot bearing.   (A12)

4. A power divider differential shows extremely high temperature damage to the interaxle differential. Technician A says a plugged oil line could cause this damage. Technician B says the driver not locking the power divider during slippery conditions could cause this damage. Who is right?
   A. A only
   B. B only
   C. Both A and B
   D. Neither A nor B   (D20)

5. A truck creeps forward from a stop when sitting with the clutch pedal depressed for a short period of time. The LEAST-Likely cause would be:
   A. a faulty master cylinder piston seal.
   B. a minute hydraulic line leak.
   C. a binding clutch linkage.
   D. a leaking or weak air servo cylinder.   (A3)

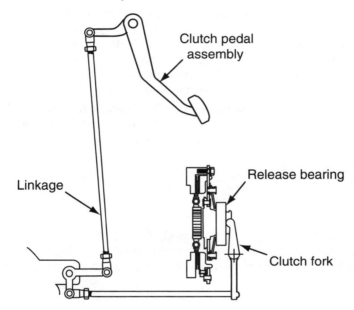

6. A clutch and linkage system as shown in the above diagram is being adjusted. Technician A says that clutch pedal free play must be adjusted first. Technician B says that the release bearing to clutch brake clearance must be performed first. Who is right?
   A. A only
   B. B only
   C. Both A and B
   D. Neither A nor B   (A5)

7. A transmission is in the shop for a rebuild. Technician A says that in addition to the rebuild, you should flush the transmission cooler system. Technician B says that unless the damage to the transmission was caused by a cooling system fault, flushing the transmission cooler system is unnecessary. Who is right?
   A. A only
   B. B only
   C. Both A and B
   D. Neither A nor B   (B20)

8. A tandem axle truck with the power divider lockout engaged has power applied to the forward rear drive axle while no power is applied to the rearward rear drive axle. The Most-Likely cause of the malfunction is:
   A. broken teeth of the forward drive axle ring gear.
   B. broken teeth of the rear drive axle ring gear.
   C. stripped output shaft splines.
   D. damaged interaxle differential.   (D17)

9. When checking transmission fluid level on a manual transmission, what is the proper procedure?
   A. Follow the guidelines stamped on the transmission dipstick.
   B. Check for proper oil level by using your finger to feel for oil through the filler plughole.
   C. Make sure that the oil level is even with the plughole.
   D. Check for proper fluid level in the transmission oil cooler sight glass.   (B11)

10. A truck has a backup light that does not turn off. All of the following are causes **EXCEPT:**
    A. a short to ground on the negative wire to the lamp.
    B. a short to positive on the negative wire to the lamp.
    C. a backup switch stuck in the open position.
    D. a short to ground on the positive wire to the lamp.   (B23)

11. When discussing wheel bearing service Technician A says that wheel bearings should always be replaced when a wheel is removed. Technician B says that raising the opposite side of the axle is a good way of filling the bearing cavity with axle lube. Who is right?
    A. A only
    B. B only
    C. Both A and B
    D. Neither A nor B   (D24)

12. Universal joint noise is being discussed. Technician A says that brinelling of the universal joint trunnions can cause noisy operation. Technician B says that a bearing with a fracture between the bearing and the bearing plate can generate noise. Who is right?
    A. A only
    B. B only
    C. Both A and B
    D. Neither A nor B   (C1)

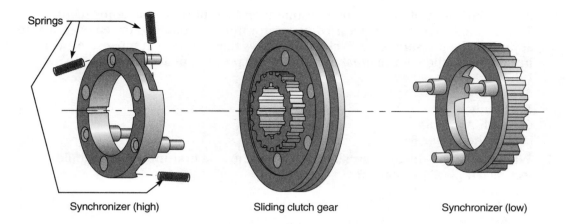

Springs

Synchronizer (high)   Sliding clutch gear   Synchronizer (low)

13. A pin synchronizer as shown in the figure is not providing adequate synchronization. Technician A says that the cone surfaces should be checked for wear. Technician B says that this condition can be caused by the driver not using the clutch while shifting. Who is right?
   A.   A only
   B.   B only
   C.   Both A and B
   D.   Neither A nor B   (B18)

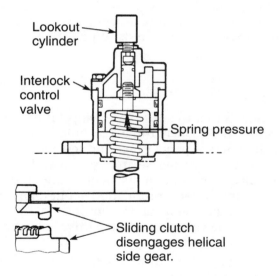

Lookout
cylinder

Interlock
control
valve

Spring pressure

Sliding clutch
disengages helical
side gear.

14. Technician A says that an air shift unit, as shown in the figure, with a sticking shift shaft can be cleaned and reinstalled. Technician B says that the air shift unit is not serviceable and should be replaced. Who is right?
   A.   A only
   B.   B only
   C.   Both A and B
   D.   Neither A nor B   (D16)

15. Driveline angles have been measured. Technician A says that the driveshaft U-joint working angles should be within one degree of each other. Technician B says that anything within three degrees of each other is acceptable. Who is right?
   A.   A only
   B.   B only
   C.   Both A and B
   D.   Neither A nor B   (C4)

16. A self-adjusting clutch is out of adjustment and requires manual adjustment. Technician A says that the wear compensator must be loosened or removed Before adjusting the pressure plate. Technician B says that the wear compensator while installed will allow for manual rotation of the pressure plate adjusting ring.
Who is right?
A. A only
B. B only
C. Both A and B
D. Neither A nor B   (A8)

17. While removing the differential carrier from the axle housing, where should the technician place jackstands?
A. Under the spring seats
B. The outermost surface of the axle housing
C. As close as possible to the axle housing cover
D. Under the brake chamber mounting bracket   (D4)

18. A transmission jumps out of gear while traveling down the road. The following are all possible causes **EXCEPT:**
A. bearings.
B. detents.
C. broken gear teeth.
D. engine mounts.   (B2)

19. When diagnosing an automated mechanical transmission, Technician A says that data link communication can only be verified by using a special data link tester. Technician B says that the data link tester or a DVOM (digital, volt, ohmmeter) can be used for harness continuity tests. Who is right?
A. A only
B. B only
C. Both A and B
D. Neither A nor B   (B8)

20. A truck driver suspects that the drive axle temperature is not accurate. Which of the following is the LEAST-Likely thing a technician would do first?
A. Remove the instrument panel gauge and test for proper movement.
B. Disconnect the drive axle temperature sensor and substitute with a variable resistance to check for proper movement of the needle.
C. Consult the manufacturer's information about temperature to resistance correlation for the axle temperature sensor.
D. Clean and grease the connection at the drive axle and retest for accuracy of the gauge.   (D25)

21. A technician notices missing teeth on the flywheel ring gear. The LEAST-Likely method to repair the problem is to:
A. replace the entire flywheel.
B. remove the flywheel and install a new ring gear.
C. use a Mig welder to replace the missing teeth.
D. send the flywheel out to a jobber for repair.   (A11)

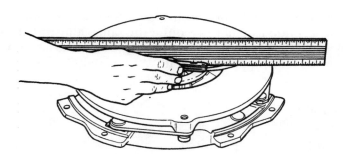

22. In the figure shown, the technician is making this measurement because the vehicle likely exhibited the following:
    A. drivetrain vibration at speeds above 35 mph.
    B. drivetrain vibration at speeds below 35 mph.
    C. chatter every time gears are shifted.
    D. chatter only at takeoff. (A5 and A6)

23. Final drive noise is heard only when cornering. Technician A says that this is common when the pinion preload is excessive. Technician B says that the problem is in the differential gearing. Who is right?
    A. A only
    B. B only
    C. Both A and B
    D. Neither A nor B (D5)

24. When checking an output shaft for excessive wear of the thrust washers, what is a common axial clearance measurement of free play?
    A. 0.005 to 0.012 inch (0.127 to 0.305 mm)
    B. 0.0065 to 0.025 inch (0.165 to 0.381 mm)
    C. 0.05 to 0.12 inch (1.27 to 3.05 mm)
    D. 0.00065 to 0.0012 inch (0.0165 to 0.0305 mm) (B16)

25. A technician notices a slight twist in the main shaft of a twin countershaft transmission during disassembly. What could cause this condition?
    A. Excessive clutch brake usage
    B. Incorrectly timed countershafts
    C. Towing the truck with the axles in place
    D. Excessive shock loading (B14)

26. A vehicle has a burnt friction disc in the clutch. Technician A says it could be caused by too much clutch pedal free play. Technician B says binding linkage may have caused it. Who is right?
    A. A only
    B. B only
    C. Both A and B
    D. Neither A nor B (A1)

27. A bearing plate style of universal joint is to be replaced. Technician A says that supporting the cross in a vice and striking the yoke with a hammer can easily remove most joints. Technician B says that the use of an appropriate puller is the recommended procedure for joint removal. Who is right?
    A. A only
    B. B only
    C. Both A and B
    D. Neither A nor B (C2)

28. A truck is being fitted with new unitized wheel hub assemblies. Technician A says that these wheel hub assemblies require the same adjustment procedures as the individual wheel bearings. Technician B says that these assemblies only require a specified torque for proper adjustment. Who is right?
    A. A only
    B. B only
    C. Both A and B
    D. Neither A nor B   (D28)

29. When adjusting a push type clutch assembly, what release bearing clearance is required?
    A. ⅛ inch (3.175 mm) between the release bearing and the clutch release levers.
    B. ½ inch (12.7 mm) between the release bearing and the clutch release levers.
    C. ⅛ inch (3.175 mm) between the release bearing and the clutch brake.
    D. ½ inch (12.7 mm) between the release bearing and the clutch brake.   (A5)

30. The LEAST-Likely cause of PTO driveshaft vibration is:
    A. a loose end yoke.
    B. an out-of-balance driveshaft.
    C. radial play in the slip spline.
    D. a slightly bent shaft tube.   (B22)

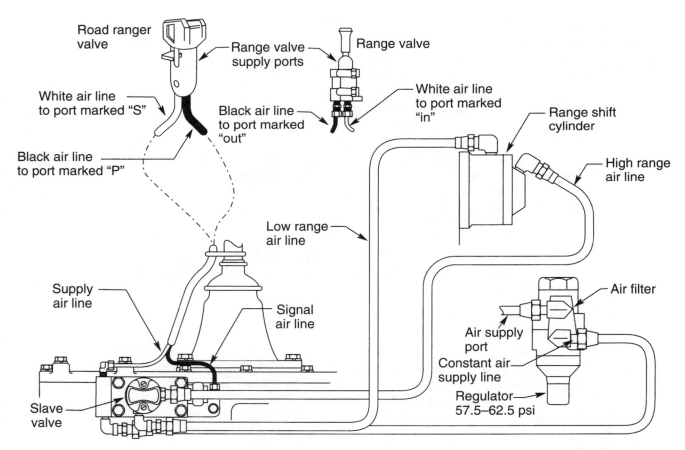

31. A truck equipped with a pneumatic high/low shift system shown in the figure will not shift into high range. What may be the cause?
    A. A dirty or plugged air filter
    B. A blown fuse
    C. A worn synchronizer in the auxiliary portion of the transmission
    D. Worn gear teeth   (B4)

32. A transmission temperature gauge does not operate. Technician A says that the sensor should be replaced because they see the same problem recur frequently. Technician B reads a value of one volt at the electrical connector for the sensor and replaces the sensor because the wiring must be intact. Who is right?
    A. A only
    B. B only
    C. Both A and B
    D. Neither A nor B   (B24)

33. A transmission has obvious signs of a leaking output shaft seal on a newer gravel hauler with only 25,000 miles on the odometer. The likely cause for the leakage is:
    A. excessive bearing wear.
    B. naturally occurring evaporation.
    C. a plugged transmission breather filter.
    D. a poor quality OEM fluid filter.   (B19)

34. A truck is experiencing driveline vibrations after having rear spring suspension work performed. Technician A says that loose U-bolts may be the cause of the vibrations. Technician B says that incorrect installation of the axle shims could cause vibrations. Who is right?
    A. A only
    B. B only
    C. Both A and B
    D. Neither A nor B   (C4)

35. A two-piece driveshaft has been removed from the truck for U-joint service. Technician A says that because of balance weights on each piece, they have to be marked for identical reassembly. Technician B says that if one of the weights gets knocked off during the U-joint replacement, the shaft will vibrate. Who is right?
    A. A only
    B. B only
    C. Both A and B
    D. Neither A nor B   (C2)

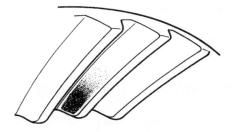

36. A technician is checking tooth contact pattern on new gears as shown in the figure. Technician A says that the ring and pinion needs to be readjusted because the contact pattern is too close to the root. Technician B says the ring and pinion needs to be readjusted because the contact pattern is too close to the toe. Who is right?
    A. A only
    B. B only
    C. Both A and B
    D. Neither A nor B   (D13)

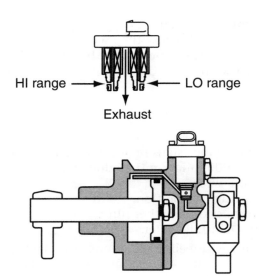

37. An electronically automated mechanical transmission is equipped with range solenoids as shown in the figure. At rest or with no voltage applied to these solenoids, what position will the range cylinder piston be in?
    A. A neutral position.
    B. Low range position.
    C. High range position.
    D. The position it was in before stopping.   (B5)

38. A burnt pressure plate may be caused by all of the following **EXCEPT:**
    A. oil on the friction disc.
    B. not enough clutch pedal free play.
    C. binding linkage.
    D. a damaged pilot bearing.   (A1)

39. A driver of a truck with a nonsynchronized transmission depresses the clutch pedal to the floor on each shift. What component is he Most-Likely to damage?
    A. Collar clutches
    B. Input shaft
    C. Clutch linkage
    D. Clutch brake   (A7)

40. What should a technician do when replacing a support bearing assembly?
    A. Lubricate the bearing.
    B. Apply lubricant to the outer bearing race to help press the bearing into place.
    C. Fill the entire cavity around the bearing with grease.
    D. Pay close attention to proper orientation of the support bearing assembly.
                                                                                  (C3)

41. A differential shows spalling on the teeth of the ring gear, while the gear teeth of the pinion have no signs of wear. What should the technician do?
    A. Replace the ring gear and pinion.
    B. Replace the ring gear.
    C. Properly rinse out the differential housing and switch to a higher viscosity of lubricant.
    D. Correctly adjust the ring gear and pinion backlash before any further damage takes place.   (D9)

42. You are setting the correct thrust screw tension on a differential that has a thrust block. Technician A says you turn the thrust screw until it stops against the ring gear or thrust block, then tighten one-half turn, and lock the jam nut. Technician B says you turn the thrust screw until it stops against the ring gear, then loosen one turn, and lock the jam nut. Who is right?
   A. A only
   B. B only
   C. Both A and B
   D. Neither A nor B   (D14)

43. A single axle truck has a noise coming from the final drive that is most pronounced on deceleration. What could cause this noise?
   A. A defective inner bearing on the drive pinion
   B. Defective differential side bearings
   C. Defective differential case gears
   D. A defective outer bearing on the drive pinion   (D1)

44. After a technician rebuilt a standard transmission, he was able to select two gears at the same time. What would allow this to happen?
   A. A sticky shift collar
   B. An interlock pin or ball left out
   C. A broken detent spring
   D. An incorrectly installed shifter lever   (B12)

45. A truck has a nonfunctional speedometer while the odometer operates correctly. The LEAST-Likely cause of the problem would be:
   A. a broken road speed sensor.
   B. loose wiring at the instrument cluster.
   C. a broken speedometer gauge.
   D. an open circuit in the wiring behind the instrument cluster.   (B21)

46. To install a new pull-type clutch a technician will need to do all the following, **EXCEPT:**
   A. align the clutch disc.
   B. adjust the self-adjusting release bearing.
   C. resurface the limited torque clutch brake.
   D. lubricate the pilot bearing.   (A6)

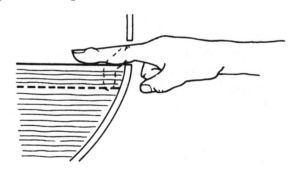

47. Technician A says the proper fluid level in the figure shown is when you can feel the lubricant with your finger. Technician B says the level must be even with the bottom of the fill hole. Who is right?
   A. A only
   B. B only
   C. Both A and B
   D. Neither A nor B   (D3)

48. Technician A says that to correctly adjust a pull type clutch requires ½ inch (1.25cm) of clearance between the release bearing and the clutch brake disc. Technician B says that to correctly adjust a pull type clutch the release fork needs ⅛ inch (0.3125cm) clearance between itself and the release bearing. Who is correct?
   A. A only
   B. B only
   C. Both A and B
   D. Neither A nor B   (A2 and A4)

49. A technician notices overheated oil coating the seals of the transmission. Technician A says that you must replace all the seals in the transmission. Technician B says that changing to a higher grade of transmission oil may be all that will be necessary. Who is right?
   A. A only
   B. B only
   C. Both A and B
   D. Neither A nor B   (B11)

50. A driver complains of a slight growl on acceleration and a buzz or tingling sensation in the gearshift lever when coasting. Technician A says that by measuring the driveline angles and entering the readings in a driveline angle analyzer program would prove that the condition is caused by the driveline. Technician B says that the condition is a transmission related condition. Who is right?
   A. A only
   B. B only
   C. Both A and B
   D. Neither A nor B   (C5)

51. A clutch pedal of an air-operated clutch is obstructed and unable to be fully depressed. Technician A says that this will allow a constant flow of air to the clutch brake control valve preventing it from operating. Technician B says that this will prevent the flow of air to the clutch brake control valve preventing it from operating. Who is right?
   A. A only
   B. B only
   C. Both A and B
   D. Neither A nor B   (A13)

52. A truck with a hydraulic retarder is in the shop with no high speed retarder operation. Technician A says that the problem is probably an air or hydraulic control circuit problem. Technician B says that the problem is probably due to a rotor failure. Who is right?
   A. A only
   B. B only
   C. Both A and B
   D. Neither A nor B   (C6)

53. A truck has a broken intermediate plate. Technician A says a broken intermediate plate can be caused by poor driver technique. Technician B says that a truck pulling loads that are too heavy can cause a broken intermediate plate. Who is right?
   A. A only
   B. B only
   C. Both A and B
   D. Neither A nor B   (A1)

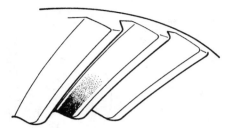

54. In the figure shown, what will the technician need to do?
    A. Adjust the pinion bearing cage shim pack.
    B. Continue assembly, the adjustment is correct.
    C. Adjust the bearing adjusting rings to decrease backlash.
    D. Adjust the bearing adjusting rings to increase backlash.   (D13)

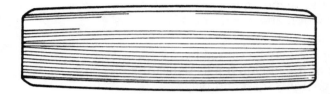

55. A technician is dismantling a transmission countershaft and notices that a bearing outer race is slightly marred as shown in the figure. What could cause this type of marking?
    A. Dirty transmission fluid
    B. Normal vibration of the transmission
    C. The bearing has been "spun"
    D. Poorly manufactured bearings   (B15)

56. The following are all acceptable steps to prepare the vehicle for a driveline angle measurement **EXCEPT:**
    A. equalize the tire pressure in all of the tires on the vehicle.
    B. if a level surface is not available to park the truck on, use jackstands to level the vehicle.
    C. jack up one of the rear tires and rotate the tire by hand until the output yoke of the transmission is vertical, then lower the vehicle.
    D. place the transmission in neutral and block the front tires.   (C4)

57. Technician A says the best way to test for a binding or stuck shift linkage is to shift between gear positions with the truck standing still. If there is any resistance while shifting into gear, the shift linkage is binding. Technician B says you have to disconnect the linkage at the transmission and check the linkage inside the transmission separately from checking the linkage outside the transmission. Who is right?
    A. A only
    B. B only
    C. Both A and B
    D. Neither A nor B   (B3)

58. A damaged pilot bearing may cause a rattling or growling noise when:
    A. the engine is idling and the clutch pedal is fully depressed and clutch released.
    B. the vehicle is decelerating in high gear with the clutch pedal released and clutch engaged.
    C. the vehicle is accelerating in low gear with the clutch pedal released and clutch engaged.
    D. the engine is idling, the transmission is in neutral, and the clutch pedal is released and clutch engaged.   (A9)

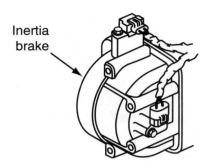

Inertia brake

59. An inertia brake as shown in the figure has a properly functioning coil but it does not slow the transmission countershafts when energized. What could cause this condition?
   A. A faulty air supply
   B. A defective diaphragm
   C. Worn friction and reaction discs
   D. A leaking accumulator   (B26)

60. Clutch slippage may be caused by:
   A. a worn or rough clutch release bearing.
   B. excessive input shaft end play.
   C. a leaking rear main seal.
   D. a weak or broken torsional springs.   (A1)

6 Additional Test Questions for Practice

Additional Test Questions

Please note the letter and number in parentheses following each question. They match the overview in Section 4 that discusses the relevant subject matter. You may want to refer to the overview using this cross-referencing key to help with questions posing problems for you.

1. A transmission temperature sensor rises five minutes after shutting the rig down. This is an indication of:
   A. nothing unusual; it is normal.
   B. a restricted pump cooling circuit.
   C. plugged or blocked transmission cooler fins.
   D. defective fluid that has lost its thermal inertia. (B24)

2. A crankshaft rear main seal is removed due to an oil contaminated clutch assembly. The technician noticed that the seal has cut a groove in the crankshaft's sealing surface. Technician A says that a wear sleeve and a matching seal should be installed. Technician B says that on some engines a thorough cleaning and a deeper installation of a standard seal is all that is required. Who is right?
   A. A only
   B. B only
   C. Both A and B
   D. Neither A nor B (A10)

3. A technician has an overheating problem on both axles of a tandem drive axle configuration. All the following could be causes **EXCEPT:**
   A. poor quality axle fluid.
   B. wheel bearings.
   C. the interaxle differential not operating smoothly.
   D. continuous vehicle overloading. (D1)

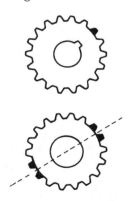

4. In the figure shown, the teeth on the gears are marked for what reason?
   A. To show worn teeth
   B. To time the gears
   C. To count the number of teeth
   D. To identify the gears (B14)

5. While discussing a twin disc clutch system, Technician A says this system is used in high torque applications. Technician B says that a twin disc clutch system uses an intermediate plate. Who is right?
   A. A only
   B. B only
   C. Both A and B
   D. Neither A nor B   (A6)

6. Technician A says drive pinion depth should be set once you properly preload the pinion bearing cage. Technician B says that setting the drive pinion depth requires adjustment of the ring gear. Who is right?
   A. A only
   B. B only
   C. Both A and B
   D. Neither A nor B   (D11)

7. A limited torque clutch brake is being used with a non-synchronized transmission. This clutch brake would LEAST-Likely be used for:
   A. slowing or stopping the input shaft when shifting into first or reverse gear.
   B. reducing gear clash when shifting from gear to gear.
   C. reducing gear damage.
   D. reducing U-joint wear.   (A7)

8. A transmission is being disassembled. All of the bearings show signs of flaking or spalling. Technician A says that this is normally caused by fatigue. Technician B says that these markings are an indication of dirt in the oil. Who is correct?
   A. A only
   B. B only
   C. Both A and B
   D. Neither A nor B   (B1)

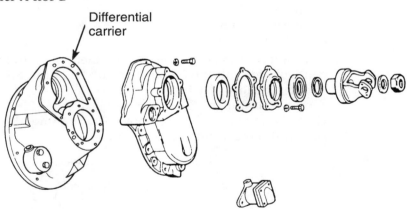

9. To remove a side gear from the power divider in the figure shown, the technician must:
   A. remove the power divider cover and all applicable gears as an assembly.
   B. disconnect the air line; remove the output and input shaft yokes, power divider cover, and all applicable gears as an assembly.
   C. remove the differential carrier and separate the differential gears from the power divider gears.
   D. remove the power divider cover and begin disassembling and separating the gears of the power divider.   (D17)

10. Because idler gears are in constant mesh they are most susceptible to:
    A. gear teeth wear.
    B. idler gear inner race spalling.
    C. idler shaft bearing wear.
    D. gear tooth chipping or breakage.   (B16)

11. While inspecting a transmission for leaks, the technician notices that the gaskets appear to be blown out of their mating sealing surfaces. What should the technician check first?
    A. The transmission breather
    B. The shifter cover
    C. The release bearing
    D. The rear seal   (B10)

12. During a routine drive axle oil change, a technician notices a few metal particles on the magnetic plug of the drive axle. What should the technician do?
    A. Inform the customer that further investigation is needed.
    B. Inform the customer of the condition and tell them to monitor the amount of particles.
    C. Do not follow up with the customer, since some metal particles are normal.
    D. Begin to disassemble the drive axle to find the cause.   (D2)

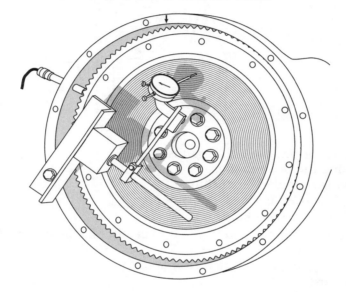

13. In the figure shown, the technician is checking for:
    A. flywheel to housing runout.
    B. flywheel face runout.
    C. flywheel radial runout.
    D. pilot bearing bore runout.   (A11)

14. A driver complains of a transmission PTO vibration only when the vehicle is shifting gears at low road speed. The vibration is likely caused by:
    A. broken gear teeth inside the PTO.
    B. a stiff or frozen U-joint in the PTO shaft.
    C. something else; the driver's diagnosis cannot be correct.
    D. bent or out of balance PTO driveshaft.   (B22)

15. When measuring driveline angles, Technician A says that the driveline angle measurement is the angle formed between the rear axle pinion shaft centerline and a true horizontal. Technician B says that the driveline angle measurement is the angle formed between the transmission output shaft centerline and the driveshaft centerline. Who is right?
    A. A only
    B. B only
    C. Both A and B
    D. Neither A nor B   (C4)

16. The best way to clean a transmission housing breather is to:
    A. replace the breather.
    B. soak the breather in gasoline.
    C. use solvent, then compressed air.
    D. use a rag to wipe the orifice clean.   (B19)

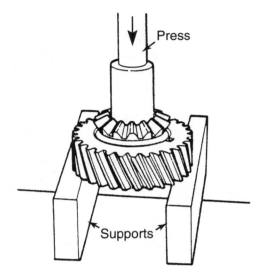

17. In the figure shown, the technician is:
    A. installing the side bearing.
    B. removing the side gear bushing.
    C. setting the bearing race.
    D. adjusting the preload.   (D5)

18. When a self-adjusting clutch is found to be out of adjustment, check all of the
    following **EXCEPT:**
    A. correct placement of the actuator arm.
    B. bent adjuster arm.
    C. frozen adjusting ring.
    D. worn pilot bearing.   (A8)

19. When discussing the adjustment of a single disc push type clutch, Technician A
    says that the clutch adjustment is within the pressure plate. Technician B says that
    the adjustment is through the linkage only. Who is right?
    A. A only
    B. B only
    C. Both A an B
    D. Neither A nor B   (A5)

20. Technician A says that lubricating slip splines requires special lithium-based
    grease. Technician B says good quality U-joint grease can also be used on slip
    splines. Who is right?
    A. A only
    B. B only
    C. Both A and B
    D. Neither A nor B   (C2)

21. To remove an oil pump from an automatic transmission, a technician must:
    A. remove the transmission, then the torque converter, then the oil pump.
    B. remove the transmission pan and filter, then remove the oil pump.
    C. remove the transmission pan and filter, then remove the main control valve body, then the oil pump.
    D. remove the transmission, then remove the torque converter, then remove the bell housing, then remove the oil pump.   (B20)

22. A technician notices a whitish milky substance when changing the fluid in an axle. This evidence of water is likely caused by:
    A. normal condensation.
    B. infrequent driving and short trips.
    C. the axle being submerged in water.
    D. the vehicle being frequently driven during rainy or wet conditions.   (D3)

23. If a vehicle has an out-of-balance driveshaft, when would symptoms likely be the most noticeable?
    A. Between 5 and 15 mph (8 and 25 kph) with no load
    B. Between 15 and 30 mph (25 and 50 kph) under load
    C. Between 30 and 45 mph (50 and 70 kph) with no load
    D. Above 50 mph (80 kph) with no load   (C1)

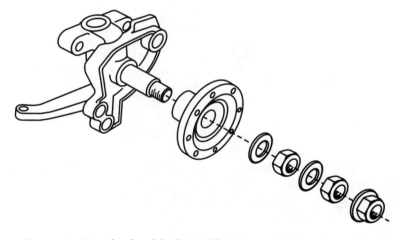

24. When inspecting a unitized wheel hub and bearing unit as shown in the figure, all of the following conditions would warrant replacement **EXCEPT:**
    A. Roughness in the bearings.
    B. Noisy operation.
    C. Bearing play of 0.003 inch.
    D. A leaking wheel seal.   (D28)

25. A conventional three-shaft drop box style of transfer case shows signs of extreme heat damage only to the input gears. The likely cause for this is:
    A. poor quality bearings.
    B. poor quality lubricant.
    C. inferior quality input gears.
    D. continuous overloading of the drivetrain.   (B24)

26. A lip seal and wiper ring are being replaced on a truck axle housing and wheel hub. Technician A says that the wiper ring should be installed with a thin coat of sealant. Technician B says when using a wiper ring, an oversized seal must be used. Who is right?
    A. A only
    B. B only
    C. Both A and B
    D. Neither A nor B   (D24)

27. Which of the following statements applies to almost all air systems?
   A. The first sign of moisture in a line indicates that repair is needed.
   B. Air flow is designed to move in only one direction so movement in the opposite direction requires mechanical means.
   C. Closed air systems do not require filters.
   D. Air circuits should not have more than three 90-degree bends per 5-foot section.   (B4)

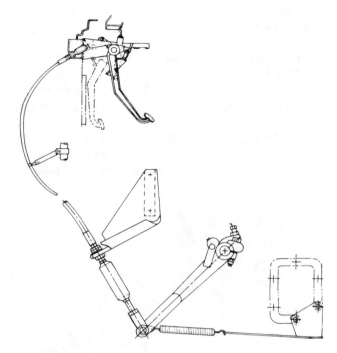

28. When adjusting a clutch linkage as shown in the figure, Technician A says that the pedal free travel should be about 1.5 to 2 inches (38.1 to 50.8 mm). Technician B says that release bearing to clutch brake travel should be less than 0.5 inch (12.7 mm). Who is right?
   A. A only
   B. B only
   C. Both A and B
   D. Neither A nor B   (A2 and A4)

29. The following are all reasons to replace an axle shaft **EXCEPT:**
   A. minute surface cracks in the axle shaft.
   B. a bent axle shaft.
   C. pitting of the axle shaft.
   D. twisting of the axle shaft.   (D21)

30. A vehicle with pre-set hub is in the shop for service. Technician A says that the hubs adjusting nut only requires a torque to specification procedure without backing off. Technician B says that the hub operates with minimal free play. Who is right?
   A. A only
   B. B only
   C. Both A and B
   D. Neither A nor B   (D28)

31. What is the most common cause of bearing failure in a transmission?
    A. Extended high torque situations
    B. Dirt in the lubricant
    C. Operating machinery in high temperature situations
    D. Poor quality of lubricant   (B11)

32. The LEAST-Likely damage to check for on an input shaft is:
    A. cracking of the pilot bearing stub.
    B. gear teeth damage.
    C. input spline damage.
    D. cracking or other fatigue wear to the input shaft spline.   (B13)

33. When diagnosing an electronically controlled, automated mechanical transmission, which tool would be LEAST-Likely used?
    A. A digital volt, ohmmeter
    B. A laptop computer
    C. A hand-held scan tool
    D. A test light   (B7)

34. A drive axle housing mating surface is slightly gouged. Technician A says a proper repair job requires you to use a torch to fill the gouges and then file them smooth. Technician B says a proper job requires you to grind and sand smooth any imperfections. Who is right?
    A. A only
    B. B only
    C. Both A and B
    D. Neither A nor B   (D19)

35. When inspecting the operation of transmission linkage, a technician should inspect it for:
    A. binding bushings.
    B. linkage length.
    C. bends and twists.
    D. surface rust and pitting.   (B3)

36. A twin countershaft transmission is being rebuilt. The lower front countershaft mounting bearing bore shows signs of scoring. Technician A says that the cause of this could be dirt or small metal particles passing through the bearing until it seized and spun. Technician B says that this could be caused by lack of lubrication due to low fluid levels that resulted in bearing seizure. Who is right?
    A. A only
    B. B only
    C. Both A and B
    D. Neither A nor B   (B1)

37. When replacing driveshaft center support bearings, a technician should always:
    A. measure any shims during removal of the old bearing.
    B. pack the bearing full of grease.
    C. measure driveline angles once the new component is installed.
    D. use hand tools because air tools could twist the bearing mounting cage.   (C3)

38. In electronically automated mechanical transmissions, an output shaft sensor sets a fault code. Technician A says that the fault code can be displayed by the service light on the dash. Technician B says that the fault code is retrievable by a hand-held scan tool. Who is right?
    A. A only
    B. B only
    C. Both A and B
    D. Neither A nor B   (B5)

39. What would be the LEAST-Likely cause of persistent oil leaks from a drive axle housing mating surface and its wheel seals?
    A. Poor quality gaskets and sealants
    B. A plugged axle housing vent
    C. A high lubricant level
    D. Incorrect axle lubricant   (D2)

40. Signs of flywheel housing mating surface wear are:
    A. a smooth dull surface texture change.
    B. gouges or other abrupt markings on the mating surface.
    C. fine hairline imperfections in the surface.
    D. pitted or additional light surface rust.   (A12)

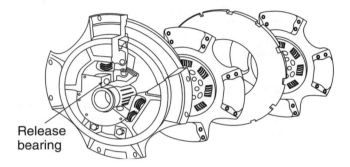

Release bearing

41. When discussing the clutch assembly shown in the figure, Technician A says that the clutch adjustment is within the pressure plate. Technician B says that this clutch would be mostly found in medium duty vehicles. Who is right?
    A. A only
    B. B only
    C. Both A and B
    D. Neither A nor B   (A5)

42. Technician A says that every time you remove a hub from an oil-lubricated type axle bearing you should repack the bearing with grease. Technician B says that every time you remove a hub from a grease-lubricated type axle bearing you should repack the bearing with grease. Who is right?
    A. A only
    B. B only
    C. Both A and B
    D. Neither A nor B   (D22)

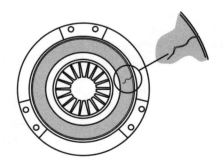

43. An intermediate plate as shown in the figure shows cracks in the surface on only one side. All of the following could be causes **EXCEPT:**
    A. release bearing that is not moving freely.
    B. poorly manufactured friction disc.
    C. friction disc that binds in worn input shaft splines.
    D. an intermediate plate that binds in the clutch cover or pot flywheel.   (A6)

44. Technician A says that the clutch teeth on a gear should have a beveled edge. Technician B says that if the clutch teeth on a gear are worn, it could cause the transmission to slip out of gear. Who is right?
    A. A only
    B. B only
    C. Both A and B
    D. Neither A nor B (B14)

45. Technician A says that if the detents in the shift tower are not aligned it could cause clutch wear. Technician B says that a broken detent spring in the shifting tower will cause the transmission to jump out of gear. Who is right?
    A. A only
    B. B only
    C. Both A and B
    D. Neither A nor B (B12)

46. What could cause a transfer case to release drive to the front axle when under load?
    A. Constant drive to the rear wheels only in high range
    B. Constant drive to the rear wheels only in low range
    C. A misadjusted range shifter
    D. Worn teeth on the front axle de-clutch (B25)

47. A technician notices excessive end play in the differential side pinion gears. How should the technician repair the problem?
    A. Split the differential case and replace the side gear pinion thrust washers.
    B. Split the differential case and replace the side pinion gears.
    C. Split the differential case and replace the side pinion gears and thrust washers.
    D. Loosen the differential side pinion gear retaining caps and install new thrust washers. (D5)

48. A technician is checking the fluid condition of an automatic transmission. What would the LEAST-Likely fluid check be?
    A. check for proper color.
    B. check for proper smell.
    C. check for any sign of particles.
    D. check for correct viscosity. (B11)

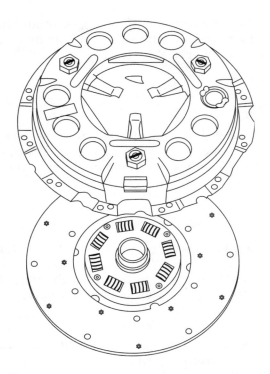

49. In the figure shown, Technician A says that this push-type clutch can be used with a clutch brake. Technician B says that this push-type clutch is used in mostly medium and light-duty applications. Who is right?
    A. A only
    B. B only
    C. Both A and B
    D. Neither A nor B   (A4)

50. If a transmission is overfilled with transmission fluid it could cause all of the following conditions **EXCEPT:**
    A. overheating of the transmission.
    B. excessive clutch wear.
    C. leakage.
    D. excessive wear to bearings and gears.   (B11)

51. When installing and adjusting wheel-bearing ends, what should the final bearing endplay be?
    A. 0.0001 inch to 0.0005 inch
    B. 0.001 inch to 0.005 inch
    C. 0.0005 inch to 0.001 inch
    D. 0.005 inch to 0.010 inch   (D24)

52. You are removing a ring gear from the differential case. Technician A uses a hammer and chisel to remove the old rivets. Technician B uses a drill and punch to remove the rivets. Who is right?
    A. A only
    B. B only
    C. Both A and B
    D. Neither A nor B   (D9)

53. A transmission has a cracked auxiliary housing. What could cause this failure?
    A. Improper driveline set-up
    B. Worn main shaft bearings
    C. A defective auxiliary synchronizer
    D. Misalignment between the engine and transmission   (B19)

54. A driver complains that with the clutch pedal pressed all the way to the floor, the transmission will not disengage. The technician has checked the fluid in the clutch master cylinder reservoir and found it to be above the MIN mark. The LEAST-Likely cause would be:
    A. improperly adjusted linkage.
    B. out of adjustment hydraulic slave cylinder.
    C. seized pilot bearing.
    D. worn clutch disc.   (A3)

55. What can cause end yoke bore misalignment?
    A. Excessive yoke retaining nut torque
    B. Excessive driveline torque
    C. Over-tightening universal joint retaining bolts
    D. Operation with poor universal joint lubrication   (C2)

56. A pot type flywheel is being resurfaced. Technician A says that only the friction surface area must be machined. Technician B says that both the friction surface and the clutch mounting surface must be machined exactly equally. Who is right?
    A. A only
    B. B only
    C. Both A and B
    D. Neither A nor B   (A11)

57. Technician A says the wear compensator is replaceable. Technician B says the wear compensator will keep the free travel in the clutch pedal within specifications. Who is right?
    A. A only
    B. B only
    C. Both A and B
    D. Neither A nor B   (A8)

58. Technician A says that a transmission mount can be thoroughly checked while it is in the vehicle. Technician B says a worn or broken transmission mount must be replaced immediately. Who is right?
    A. A only
    B. B only
    C. Both A and B
    D. Neither A nor B   (B9)

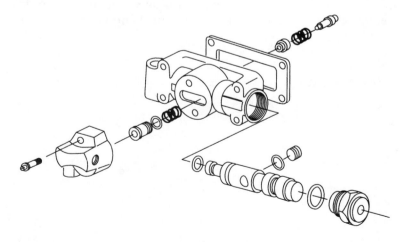

59. What type of slave valve is shown in the figure?
    A. The piston slave valve
    B. The poppet slave valve
    C. The low port slave valve
    D. The low range slave valve   (B4)

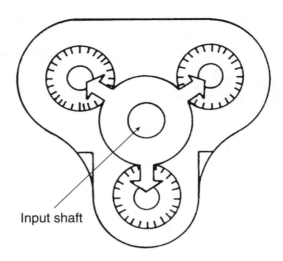

Input shaft

60. When timing a triple countershaft transmission as shown in the figure, what should the technician do?
   A. Align to timing marks visible from previous rebuilds or service.
   B. Mark the gears before disassembly, then align those marks during assembly.
   C. Align the keyway so that all countershaft keyway align with the main shaft.
   D. Align the timing tooth of each countershaft with the corresponding timing mark on the main shaft.   (B14)

61. Flywheel face runout is measured by:
   A. attaching a dial indicator to the center of the flywheel and measuring the flywheel face by turning the crankshaft.
   B. pushing the flywheel in, attaching a dial indicator to the flywheel-housing bore, and rotating the flywheel.
   C. pulling the flywheel out, attaching a dial indicator to the flywheel-housing bore, and rotating the flywheel.
   D. removing and resurfacing the flywheel if flywheel face runout is suspected.
                                                                            (A11)

62. When servicing a clutch, checking the crankshaft end play is sometimes required. Technician A says that a dial indicator must be used to measure in and out movement of the crankshaft. Technician B says that the dial indicator must measure crankshaft up and down movement. Who is correct?
   A. A only
   B. B only
   C. Both A and B
   D. Neither A nor B   (A10)

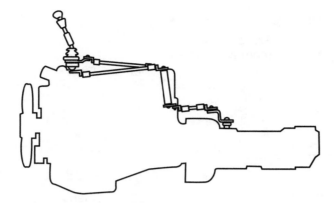

63. A vehicle with a transmission linkage system as shown in the figure is in for service. Technician A says that checking a transmission shift linkage for wear is not necessary if you can properly make all of the necessary adjustments. Technician B says that the shift linkage should always be checked. Who is right?
    A. A only
    B. B only
    C. Both A and B
    D. Neither A nor B   (B3)

64. A tractor with a locking differential will not release (unlock). What could cause this condition?
    A. A lack of air supply to the shift cylinder
    B. A broken air line to the shift cylinder
    C. Damaged teeth on the shift collar
    D. A broken shift cylinder return spring   (D6)

65. Which of the following is LEAST-Likely to cause noise in a manual transmission?
    A. A broken detent spring
    B. A worn or pitted input bearing
    C. A worn or pitted output bearing
    D. A worn countershaft bearing   (B2)

66. Technician A says that automatic transmission gaskets cannot be replaced by silicon sealants because of possible ingress into the transmission hydraulic system. Technician B says that an appropriate silicon sealant can be used to seal a porous case. Who is right?
    A. A only
    B. B only
    C. Both A and B
    D. Neither A nor B   (B9)

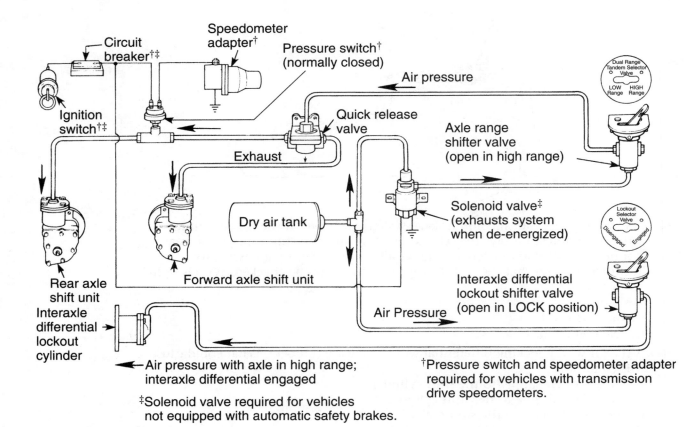

Air pressure with axle in high range; interaxle differential engaged

†Pressure switch and speedometer adapter required for vehicles with transmission drive speedometers.

‡Solenoid valve required for vehicles not equipped with automatic safety brakes.

67. This is the axle range and interaxle differential lockout schematic of a vehicle that will not shift from high range to low range. The Most-Likely cause is:
   A. a faulty air compressor.
   B. an air leak at the axle shift unit.
   C. a quick release valve.
   D. a plugged air filter.   (D16)

68. All of the following may cause premature clutch disc failure **EXCEPT:**
   A. oil contamination of the disc.
   B. worn torsion springs.
   C. worn U-joints.
   D. a worn clutch linkage.   (A1)

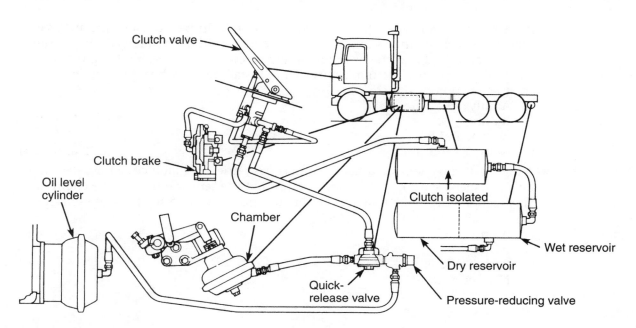

69. A truck equipped with an air-operated clutch as shown in the figure has a minute leak in the air compressor supply circuit that allows pressure loss overnight. Technician A says that this will prevent clutch brake operation during start up. Technician B says that the clutch should still be releasable during start up without air pressure in the supply circuit. Who is right?
    A. A only
    B. B only
    C. Both A and B
    D. Neither A nor B   (B22)

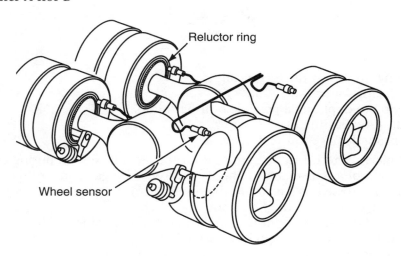

70. A wheel speed sensor as shown in the figure is being checked for proper operation with the wheel raised, the sensor disconnected and the wheel being rotated, Technician A says that the sensor output could be affected by the sensors adjustment. Technician B says that the sensor output should be checked with an AC voltmeter. Who is right?
    A. A only
    B. B only
    C. Both A and B
    D. Neither A nor B   (D26)

71. A tractors rear axle wheel hub is removed for brake inspection. Technician A says that the wheel bearings and seals should be inspected before reinstalling the hub. Technician B says that the bearings should be coated with fresh lubricant and fresh oil poured into the hub cavity before installing the wheel hub and operating the tractor. Who is right?
    A. A only
    B. B only
    C. Both A and B
    D. Neither A nor B   (D22 and D24)

72. A digital data reader (hand-held scanner) has retrieved a fault code for a defective tailshaft speed sensor. Technician A states that the sensor must now be changed. Technician B says that a digital multimeter should be used to check the sensor before replacement. Who is right?
    A. A only
    B. B only
    C. Both A and B
    D. Neither A nor B   (B5)

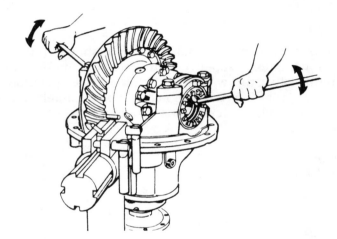

73. In the figure shown the technician is:
    A. checking for bearing wear.
    B. trying to duplicate a possible bearing sound.
    C. adjusting bearing play.
    D. causing damage to the differential.   (D12)

74. A technician is reconnecting air lines for the air shift system during the installation of a transmission. The slave valve to range cylinder lines are mistakenly crossed. What would the result of this mistake be?
    A. No range shifting could be made
    B. A constant air loss from the exhaust port of the control valve
    C. Low range air loss through the slave valve
    D. Low range operation when high range is selected   (B6)

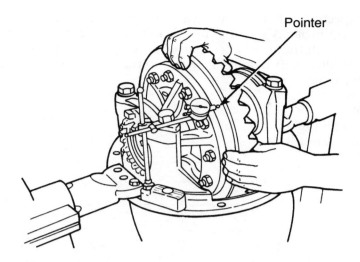

75. Technician A says that ring gear run out is measured before preloading the differential side gear bearings as shown in the figure. Technician B says that ring gear run out is adjusted out by preloading the differential side gear bearings. Who is right?
    A. A only
    B. B only
    C. Both A and B
    D. Neither A nor B   (D8)

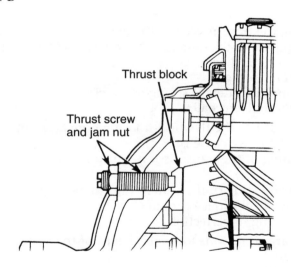

76. When discussing the thrust block shown in the figure, Technician A says that it should be installed one half turn away from the ring gear. Technician B says that it is normal to see light scoring on the thrust block. Who is correct?
    A. A only
    B. B only
    C. Both A and B
    D. Neither A nor B   (D14)

77. When inspecting a disassembled twin countershaft transmission with high mileage, Technician A says that reverse idler shaft wear is common due to the loading through the small idler gears in reverse operation. Technician B says that the reverse idler shaft wear is likely due to the separating forces during reverse operation. Who is correct?
   A. A only
   B. B only
   C. Both A and B
   D. Neither A nor B   (B17)

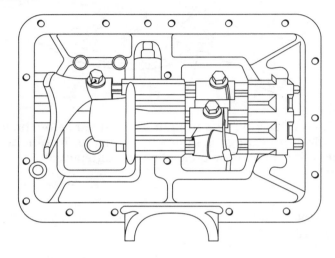

78. When checking the transmission shift cover detents on the shift bar housing as shown in the figure, check for all of the following **EXCEPT:**
   A. worn or oblonged detent recesses.
   B. broken detent springs.
   C. properly lubricated detent spring channels.
   D. rough or worn detent ball.   (B11)

79. A technician measures the flywheel-housing bore face runout to be out of specification. The likely cause is:
   A. overtightening of the transmission, causing undue pressure on the housing face.
   B. extreme overheating of the clutch, causing warpage in the flywheel housing.
   C. excessive flywheel surface runout.
   D. a manufacturing imperfection.   (A12)

80. The following are all reasons for replacement of the pilot bearing **EXCEPT:**
   A. rough action.
   B. binding action.
   C. spalled outer race.
   D. excessive end play.   (D7)

81. Technician A says that a clutch brake needs to be inspected for wear and fatigue. Technician B says that as long as the component is in place, it will function properly. Who is right?
   A. A only
   B. B only
   C. Both A and B
   D. Neither A nor B   (A7)